STUDENT'S SOLUTIONS MANUAL

SARAH STREETT

STATISTICS FOR BUSINESS:
DECISION MAKING AND ANALYSIS

Robert Stine
The Wharton School of the University of Pennsylvania

Dean Foster
The Wharton School of the University of Pennsylvania

The author and publisher of this book have used their best efforts in preparing this book. These efforts include the development, research, and testing of the theories and programs to determine their effectiveness. The author and publisher make no warranty of any kind, expressed or implied, with regard to these programs or the documentation contained in this book. The author and publisher shall not be liable in any event for incidental or consequential damages in connection with, or arising out of, the furnishing, performance, or use of these programs.

Reproduced by Pearson Addison-Wesley from electronic files supplied by the author.

Copyright © 2011 Pearson Education, Inc.
Publishing as Pearson Addison-Wesley, 75 Arlington Street, Boston, MA 02116.

All rights reserved. No part of this publication may be reproduced, stored in a retrieval system, or transmitted, in any form or by any means, electronic, mechanical, photocopying, recording, or otherwise, without the prior written permission of the publisher. Printed in the United States of America.

ISBN-13: 978-0-321-28614-7
ISBN-10: 0-321-28614-6

2 3 4 5 6 BRR 14 13 12 11

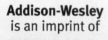

www.pearsonhighered.com

This Student's Solutions Manual includes worked out solutions to all of the odd numbered exercises in the exercise sets.

TABLE OF CONTENTS

Chapter 2 Data	1
Chapter 3 Describing Categorical Data	2
Chapter 4 Describing Numerical Data	8
Chapter 5 Association between Categorical Variables	13
Chapter 6 Association between Quantitative Variables	17
Chapter 7 Probability	22
Chapter 8 Conditional Probability	25
Chapter 9 Random Variables	28
Chapter 10 Association between Random Variables	32
Chapter 11 Probability Models for Counts	35
Chapter 12 The Normal Probability Model	37
Chapter 13 Samples and Surveys	40
Chapter 14 Sampling Variation and Quality	42
Chapter 15 Confidence Intervals	45
Chapter 16 Statistical Tests	48
Chapter 17 Alternative Approaches to Inference	51
Chapter 18 Comparison	53
Chapter 19 Linear Patterns	55
Chapter 20 Curved Patterns	60
Chapter 21 The Simple Regression Model	65
Chapter 22 Regression Diagnostics	72
Chapter 23 Multiple Regression	77
Chapter 24 Building Regression Models	86
Chapter 25 Categorical Explanatory Variables	91
Chapter 26 Analysis of Variance	100
Chapter 27 Time Series	106

Chapter 2 Data

1. Variable Name: brand of car; Type: categorical; Cases: drivers
3. Variable Name: color preference; Type: categorical; Cases: consumers in focus group
5. Variable Name: item size; Type: ordinal; Cases: unknown (could be stocks in stores or purchase amounts)
7. Variable Name: stock price; Type: numerical; Cases: companies (though the question is vague)
9. Variable Name: Sex; Type: categorical; Cases: respondents in survey

True/False

11. False. Zip codes are numbers, but it would not be sensible to average them, for example.
13. False. Cases is another name for the rows in a data table.
15. True.
17. False. A Likert scale is used for ordinal data.
19. False. Aggregation collapses a table into one with fewer rows.

Think About It

21. (a) The data is cross sectional.
    (b) The variables are Whether the employee opened an IRA (categorical) and the Amount saved (numerical with dollars as the units).
    (c) Did employees respond honestly, particularly when it came to the amount they reported to have saved?

23. (a) The data is cross sectional.
    (b) The variable is the Service rating (ordinal most likely, using a Likert scale).
    (c) With only 500 replying, are the respondents representative of the other guests?

25. (a) The data is a time series.
    (b) The variable is the Exchange rate of the US dollar to the Canadian dollar (numerical ratio of currencies).
    (c) Are the fluctuations in 2005 typical of other years?

27. (a) The data is cross-sectional.
    (b) The variables are the Quality of the graphics (categorical, perhaps ordinal from bad to good) and the Degree of violence (categorical, perhaps ordinal from none to too much).
    (c) Did some of the participants influence the opinions of others?

29. (a) The data is cross sectional (though it could be converted to a time series).
    (b) The variables are Name (categorical), Zip code (categorical), Region (categorical), Date of purchase (categorical or numerical, depending on the context; the company could compute the average length of time since the last purchase), Amount of purchase (numerical with dollars as the units) and Item purchased (categorical).
    (c) Presumably the region was recorded from the zip code.

Chapter 3 Describing Categorical Data

Mix and Match

In each case, unless noted, bar charts are better to emphasize counts whereas pie charts are better to communicate the relative share of the total amount.

1. Proportion of autos: pie chart is the most common; a bar chart or Pareto chart can also be used.
3. Coupons: bar chart or Pareto chart (these are counts) or perhaps a table (only three values)
5. Destination: bar chart or Pareto chart (counts) or pie chart (shares)
7. Excuses: Pareto chart
9. Software: pie chart (shares) or perhaps a table (only three values)
11. Ratings: Bar chart or table (only four values). Because the values are ordinal, avoid a pie chart.

True/False

13. True, but only in general. For variables with few categories, a frequency table is often better, particularly when the analysis requires knowing the detailed frequencies.
15. False. The frequency is the count of the items.
17. True. It would be false if the variable were ordinal; you should not put the shares of an ordinal variable into a pie chart.
19. True.
21. True.

Think About It

23. The message is that customers tend to stick with manufacturers from the same region. Someone trading in a domestic car tends to get another domestic car whereas someone who trades in an Asian car buys an Asian car. There's not a lot of switching of loyalties. The more subtle message, one that is disturbing to domestic car makers, is that those who own Asian cars are more loyal (78% buy another Asian car compared to 69% who stick with a domestic car). That makes it hard for domestic manufacturers to win back customers, even if they improve the quality of their cars.

25. This is a bar chart if you think about the underlying data as labeling the dollars held in these countries. The intent of the plot is to show the relative sizes of these counts, comparing the shares of U.S. debt held in these countries.

27. (a) No, these categories are not mutually exclusive. These percentages summarize four dichotomous variables, not one variable.
(b) Divided bars such as these might work well. This style is commonly used in reporting opinion poll results in the news. Sorting the values so that the percentages are in order also makes for a cleaner presentation.

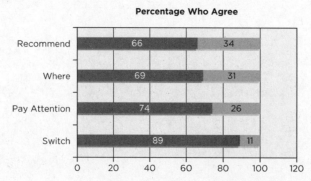

Chapter 3 Describing Categorical Data

29. No. These percentages only list the percent of executives that report each problem and are not the relative frequencies of a categorical variable. The categories are not mutually exclusive; some of the executives listed several issues.

31. This variable is continuous and most of the values would be unique, producing a bar chart with close to 200 categories, one for each purchase amount. (Of course, there might be quite a few $0.99 coffee or soda purchases.)

33. The bar chart would have one very long bar (height 900) and five shorter bars of height 20 each. The plot would not be very useful, other than to show the predominance of one category.

35. The bar chart would have five bars, each of the same height.

37. The frequency table is simple enough to look at with only two numbers, particularly if it includes proportions.

39. The mode is Public. There's no median for this chart; the data is not ordered.

41. The manufacturers want to know the modal preference because it identifies the most common color preference. Color preferences cannot be ordered, the median color preference is not defined.

43. This is ordinal data. Though not evidently a Likert scale (note the missing middle category), the responses are ordered. A pie chart would not be appropriate as it would conceal the ordering.

You Do It

45. (a) The underlying data table probably accumulates case sales by brand to some degree, such as the number sold by different retail chains and perhaps in different months. All we have here are the aggregated totals.

    (b) To show the shares, a pie chart is most natural because it implies that we are dividing up the total amount. With all the brands shown, however, the small categories make it hard to show the labels. This version accumulates the 3 smallest brands into an "other" category.

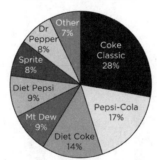

    (c) Here is the pie chart, organized by company rather than brand. If we were artists, then a nicer way to prepare this chart would be to show at the same time the division of Pepsi into brands and so forth. The 38% share of Pepsi comes from Pepsi-Cola, Mt Dew, Diet Pepsi, and Sierra Mist. These divisions could then be used to slice up Pepsi's share of the chart.

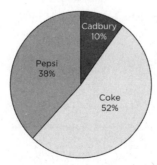

    (d) To show the amounts sold, here's a bar chart with the two accumulated types.

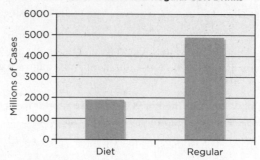

47. (a) The Other category forms an additional row in the tables so that each column adds up to 100%. The addition of this extra row makes up a big part of both pie charts.

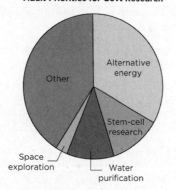
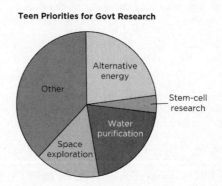

(b) The side-by-side bar chart works well for this. Notice that we no longer need the Other category that dominates the pie charts.

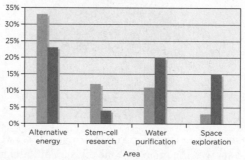

(c) No, because the categories would no longer partition the cases into distinct, non-overlapping subsets. A pie chart should only be used to summarize mutually exclusive groups.

49. It's hard to beat this combined bar chart, though some would rather see two pie charts placed side by side. From the bar chart, you can quickly tell the shares each year by the colors, and the adjacent bars show that Splenda gained shares while the others fell.

Chapter 3 Describing Categorical Data

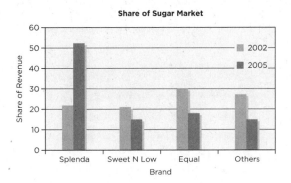

51. (a) Visually, the distributions in the two bar charts seem similar. You should not use one bar chart because the scale required to show all businesses would hide the counts for the businesses owned by women.

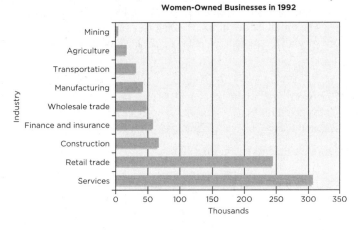

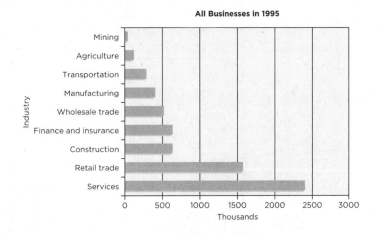

(b) A pie chart shows which industries have the most women-owned businesses, but it does not show the percentage of women-owned businesses *within* each industry. It is necessary to form a column that shows the proportion of women-owned businesses within each industry. This plot shows the percentages within each industry. While some have more and some less, women own about 10-15% across the board, with none particularly standing out.

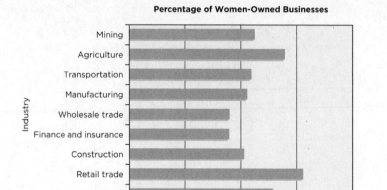

(c) Yes, but slight. The problem is that some industries might have grown or shrunk in number during the three intervening years. The broad nature of these categories suggests that any change was relatively small. If the data tracked more specialized industries, there might have been a larger proportional change over the three years.

53. (a) Use a table with the two rows and the percentages (or proportions)

| Unexpected illness | 4,463 | 15.8% |
| Planned leave | 23,735 | 84.2% |

(b) A Pareto chart shows the categories in order of size.

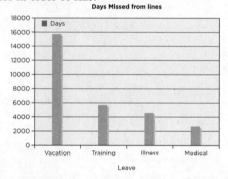

55. (a) Yes, pie charts are fine because the responses are mutually exclusive and sum to 100%.

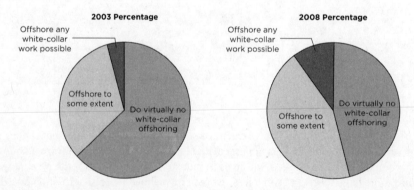

(b) Various answers are possible. The following layout is reasonable.

Chapter 3 Describing Categorical Data

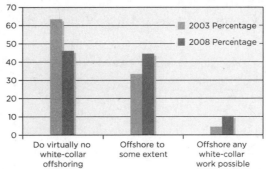

(c) The bar chart facilitates comparison. The pie chart makes the relative shares more apparent. For example, the 2003 pie shows a predominant share for taking no action, the only choice anticipated to fall in 2008.

(d) The mode and median agree (virtually none) in 2003, but differ in 2008 as responses shift from the more consistent response to a tendency to do more off shoring.

57. (a) The breakdown of employment into so many categories hides the dominance of computer-related occupations; you have to look at the labels to see that all of the top categories are related to computing. Statisticians have just about fallen off of the visible portion of the list shown (by Excel) next to the pie chart

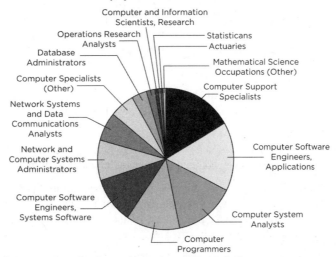

(b) The most common of these categories is computer support specialists.

(c) It makes sense to combine some of the categories. Here's one suggestion that bundles all of the computer titles into one category. Networking computers is second. Mathematics seems smaller than ever.

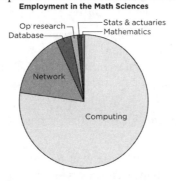

Chapter 4 Describing Numerical Data

Mix and Match

1. g
3. h
5. j
7. d
9. e

True/False

11. False. The box is the median, with its lower edge at the 25% point (lower quartile) and its upper edge at the 75% point (upper quartile).

13. True.

15. True.

17. False. The Empirical Rule applies only to numerical variables that have a symmetric, bell-shaped distribution.

19. True.

21. True. In the absence of variation, all of the data values are the same, and hence the mean and median are equal to this common value.

Think About It

23. You cannot tell from the median. There could be, for example, one very, very long song that filled the Shuffle by itself. Because this one large song does not affect the median, the median could be small but the songs would not fit.

25. The histogram of incomes in the United States is very heavily right skewed. It's hard to get very far below zero (people with negative incomes have received tax credits), but the upper limit is in the stars.

27. The payments on an adjustable rate mortgage are more variable than those on a fixed rate mortgage. If interest rates climb, then the required payments climb as well. In this context, variation is "bad" in the sense that the homeowner might see a large increase in the monthly payment if interest rates climb. Fixed rate mortgages hold the payment fixed, eliminating the variation in payment.

29. The range is very sensitive to outliers because it is the difference between the two most extreme values. The presence of large (or small) outliers increases the size of the range.

31. Mortgage payments have a larger SD because these payments are so much larger than allowances. It's likely, though, that the coefficient of variation may be larger for the allowances. The mean mortgage payment is much larger than the mean allowance!

33. (a) The group with the largest mean, music only.
    (b) No, because you cannot recover the total amount from the median.
    (c) No, because we do not know that every shopper bought the same amount; we'd expect substantial variation in the data with overlap between groups.

35. No. If the distribution is bell shaped, the Empirical Rule suggests this is a common occurrence. Even if it's not bell shaped exactly, 1 SD is not far from the mean.

37. (a) Right skewed, with a peak at zero (for those without an iPod) and tailing off to the right.
    (b) Right skewed with one mode, from moderate prices to very large orders.
    (c) Bell shaped around the target weight (or perhaps a weight above the target).
    (d) It depends, but this distribution may be bimodal if there is a mix of male and female students.

39. (a) 11. Add the heights of the two bins located between 4 and 5.
    (b) The mean is slightly less than 5% (4.88). The outliers that charge no tax (or very small) pull the mean to the

left.
(c) The median is larger. The outlier to the far left pulls down the mean, but has less impact on the median.
(d) The rates are rounded to values like 6.5 or 5.5, producing the isolated peaks.
(e) The SD is about 2. The mean is near 5%. If the SD were 5, then the mean ± 1 SD would hold all of the data. It would take a range of about ±3 SDs to make that happen.

41. (a) The mean is $34,000 compared to the median at $27,000. You can get the median from the centerline of the boxplot. You know that the mean is larger because of the skewness, but it is hard to guess how much larger. It would have to be very skewed, however, for the mean to exceed the upper quartile (which is less than $50,000).
(b) The IQR is about $30,000.
(c) Usually, the SD is smaller than the IQR (about 2/3 of the size), but the skewness and presence of outliers changes that. The two are about the same size here, but the SD is slightly larger.
(d) Only the labels on the x axis would change, dropping 3 zeros from each. Otherwise, the figure is identical.

43. There's little difference, on average, unless errors in favor of the store, for instance, happen more frequently for expensive items and errors in favor of the customer happen more frequently for inexpensive items. The pricing errors will, however, lead to more variation in prices and a more spread out histogram. For example, if every customer were to buy the same item, the standard deviation would be zero, except for the additional presence of these pricing errors.

45. The distribution of income is very right skewed, with the upper tail reaching out to very high incomes. In this case, the mean will be larger than the median.

47. The shape will be the same as in Figure 4-1; only the labels on the x axis will change. Rather than 1, 2, 3,..., you will see 60, 120, 180 and so forth. The count axis and bin heights will be the same.

49. (a) Multiply each of these summary statistics by 60, as shown below.

Summary	File Size MB	Song Length Sec
Mean	3.8	228
Median	3.5	210
IQR	1.5	90
Standard deviation	1.6	96

(b) The mean and median would increase by 2 MB, but the IQR and SD would remain the same.
(c) Such a song is longer than all of the others (see the text discussion). Hence, the median would stay about where it is (perhaps shifting up to the next larger value in the sorted list of sizes) and the IQR would remain the same.

51. (a) The mean and SD are $17,950 and $11,830, but these do not capture the bimodal nature of the data.
(b) The data are bimodal with cluster centers having means of about $7,500 and $30,000.
(c) It's hard to see, but in fact, there's a clue that the data is multi modal in the boxplot. Because the groups are of comparable size, the box of the boxplot is long relative to the size of the whiskers on either side. Unless you're looking carefully, however, you'd never recognize that this is a signal for a bimodal shape.
(d) The schools in the cluster with mean near $7,500 are public; the others are private schools.

You Do It

53. (a) The histogram (shown below) is right-skewed. Some of these cars have exceptionally high horsepower.

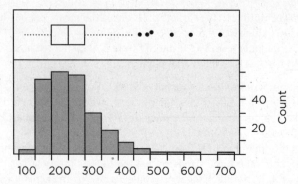

(b) The histogram conveys more about the shape of the distribution, whereas the boxplot nails down the position of the median and the IQR. The boxplot also highlights the outlying cars with very high power.

(c) The mean is 214 HP with standard deviation 84 HP. The mean is the balance point of the histogram, to the right of the center of the large peak. The SD suggests the spread of the data, but because the distribution is not bell shaped, we should not rely on the Empirical Rule in this case. (Even so, the Empirical Rule gives reasonable ranges. For example, the HP of 95% of these cars is in the range $\bar{y} \pm 2$ SD.)

(d) $c_v = 84/214 \approx 0.39$; s is about 40% of the size of the mean.

(e) The unusually powerful cars include several Ferrari models, a Lamborghini, and an Aston Martin. These have exorbitant horsepower, they are not coding errors.

(f) The median horsepower is 200, so a car with this much power is rather typical. Half of the models have more, and half have less.

55. (a) This table summarizes the distribution, with and without "Hey Jude", the large positive outlier to the right of the distribution. The calculation of the IQR may vary slightly depending on your software's calculation of the quartiles, but should be close to the values shown here.

	With Outlier	Without Outlier
Maximum	6.5525	3.9932
Quartile	2.9795	2.9619
Median	2.4654	2.4276
Quartile	2.1596	2.1543
Minimum	1.8212	1.8212
IQR	0.82	0.81
Mean	2.7343786	2.5875276
Std. Dev.	0.9503484	0.5777002
N	27	26

(b) By setting aside Hey Jude and recalculating the summary measures, we can see exactly how the outlier affects the summaries. Because it is based on percentiles, not averaging, the IQR hardly changes at all. The SD falls about 60% of the size with all 27 songs.

(c) The SD changes more, proportionally. The mean falls from 2.73 to 2.59 (95% of its prior value), but the SD drops from 0.95 down to 0.58 (about 60% of its prior value).

57. (a) There is missing data for the number of employees for 15 of the companies, leaving 181 ratios.

(b) See the table below. The values are thousands of dollars in net sales per employee (millions divided by thousands leaves thousands per employee).

(c) Unimodal, but very right skewed. The distribution is not remotely bell shaped.

(d) DynaBazaar has the largest ratio, approaching $7,000,000 in sales per employee. The ratio is so large because the data indicate only one employee (0.001 thousand = 1).

(e) First, we have to decide what it means to be comparable. We could use the mean or median level of firms in this industry, or perhaps other values. We'd probably want to stay inside the quartiles. For instance, to be at the median, a firm with 15 employees would have to generate 15 × 160,600 = $2,409,000 in sales to reach the middle of the pack.

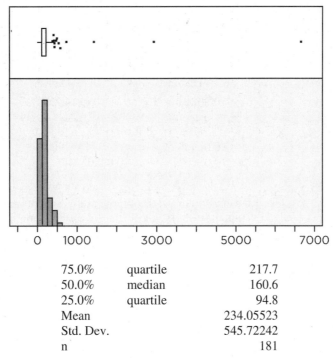

75.0%	quartile	217.7
50.0%	median	160.6
25.0%	quartile	94.8
Mean		234.05523
Std. Dev.		545.72242
n		181

59. (a) The three histograms appear below. All are bell shaped with a few outliers. Each seems to have a slightly positive mean, with more variation in the returns on Disney and McDonalds, and less variation in the returns on Exxon.

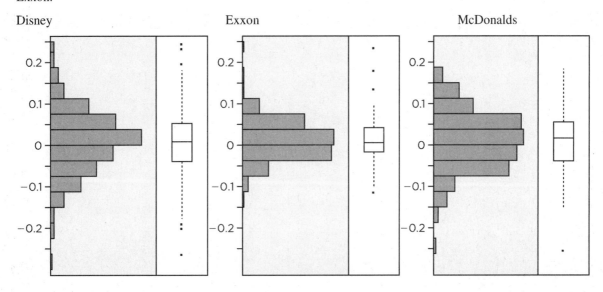

(b) The means and SDs appear below. Because of the bell shaped distributions, the Empirical Rule works well for these distributions.

	Disney	**Exxon**	**McDonalds**
Mean	0.0086841	0.0116506	0.0104956
Std. Dev.	0.0791277	0.0460873	0.0712353
Coef. of var.	9.11179	3.95579	6.78716
N	192	192	192

(c) The values of c_v indicate that returns on Disney are much more variable relative to the mean return than those on the others, then McDonalds, with Exxon the most steady. As in exercise 57, the c_v may be unreliable because the means are so close to zero.

(d) It's not true for these. Exxon has the highest mean return as well as the smallest SD. It's the best from both points of view.

61. (a) The table below shows the Sharpe ratios. Exxon's is largest, follow by McDonalds and then Disney.

	Disney	Exxon	McDonalds
Mean	0.0086841	0.0116506	0.0104956
Std. Dev.	0.0791277	0.0460873	0.0712353
Sharpe	0.0680432	0.1811909	0.1010117
Dec. 2005	-0.0276775	-0.0320524	-0.0038404

(b) The Sharpe ratio for Exxon.

(c) No. Exxon had the largest loss. Like any summary, the Sharpe ratio describes an average level of performance, but it's not going to tell you the ordering of the three for every month. There's variation in the returns, and this month, Exxon was the loser, even though on average it came out on top.

Chapter 5 Association between Categorical Variables

Mix and Match

1. j
3. e
5. a
7. i
9. d

True/False

11. True, at least in principle. If the percentages in the table match those in the margins, there's little association. The counts might not match perfectly due to rounding, but they should be very close.

13. True.

15. True.

17. False. The value of chi-square is the same if the rows and columns are reversed.

19. False. Association is not the same as cause and effect. We cannot interpret association as causation because of the possible presence of a lurking variable. Some managers may operate under very different conditions.

21. True.

Think about It

23. (a) Finding association would mean that employed and retired respondents found different rates of satisfaction in resolving the disputed charge.
    (b) The two variables are not associated. Roughly equal proportions of both groups of callers (about 70%, 697/1,000 and 715/1,000) were satisfied with the outcome of their call.

25. A lack of association makes this division much simpler, because then the choice of the best color does not affect the packaging. If the two are associated, then the preferred style of packaging depends on the color. In this case, the two divisions would have to coordinate their efforts more closely.

27. (a) The administration group. The red segment is largest for this group.
    (b) The variables are associated because the composition of the segments differs for the three groups. In particular, we can see that family issues are a more prominent reason for missing work in the administration group than the other two.

29. (a) Asia Pacific. This is the widest column in the plot (about 29% of the total).
    (b) Latin America and Middle East/Africa. The red share of these is more than half. (Use the scale at the left of the plot).
    (c) 80% of the Asia/Pacific market would be larger.
    (d) No. The figure emphasizes the share of brand within the region, not the other way around. (it's about one-quarter).
    (e) Yes. The conditional distribution depends on the location. One manufacturer dominates in one region (Japan Tobacco in Asia Pacific) whereas another is dominant elsewhere (Phillip Morris in North America).

31. If the choice of color is not associated with the style of the vehicle, then yes. Otherwise, it may be the case that the choice of color depends on the type of vehicle.

33. These are most likely associated. Children are less likely to be in stores at night.

35. Association is not the same as causation. It could be the case, as the eminent statistician R. A. Fisher argued, that both smoking and cancer share a common cause such as some underlying genetic characteristic that causes cancer and also produces a desire to smoke. His objections have not stood the test of time, and most now accept the relationship between smoking and cancer as causal.

37. (a) $V = 0$. The rows are proportional, and the two variables are not associated with each other.
(b) $V = 0$. Because it is designed to measure the association between categorical variables (which do not define an ordering), Cramer's V does not depend on the order or arrangement of the rows in the table. You can rearrange the rows and columns; the value of V remains the same.
(c) The lack of association means that when ordering paints, the store should order the same fraction of gloss in each color (4/7 low gloss, 1/7 medium, and 2/7 high gloss).

You Do It

39. (a)

	Weekday	Weekend	Total
Premium	126	62	188
Plus	103	28	131
Regular	448	115	563
Total	677	205	882

(b) Among weekday purchases, 126/677 = 19% are for premium, 103/677 = 15% are for plus, and the remaining 66% are for regular.
(c) 126/188=67% of premium purchases are on weekdays, and 33% on weekends.
(d) No. These conditional distributions are not directly comparable. One refers to the type of gas given that the purchase happens on the weekday, whereas the second refers to the timing of the purchase given that premium gas was bought. To illustrate association, one can, for example, compare the conditional distribution of purchases of regular gas to the answer in part c. For regular purchases 448/563 = 80% occur on weekdays, compared to 67% of premium purchases.
(e) Weekends. Though more premium gas is sold during the week, there's a greater concentration of premium sales on weekend days.

41. (a) The expected counts for the table are

$(188 \cdot 677)/882 = 144.30$ $(188 \cdot 205)/882 = 43.70$
$(131 \cdot 677)/882 = 100.55$ $(131 \cdot 677)/882 = 30.45$
$(563 \cdot 677)/882 = 432.14$ $(563 \cdot 677)/882 = 130.86$

and from these we arrive at the following table of contributions to the overall chi-square for the table:

$$\frac{(126-144.3)^2}{144.3}=2.32 \qquad \frac{(62-43.70)^2}{43.70}=7.67$$

$$\frac{(103-100.55)^2}{100.55}=0.06 \qquad \frac{(28-30.45)^2}{30.45}=0.20$$

$$\frac{(448-432.14)^2}{432.14}=0.58 \qquad \frac{(115-130.86)^2}{130.86}=1.92$$

Summing these gives $\chi^2 = 12.75$ and $V = \sqrt{\frac{12.75}{882 \cdot 1}} = 0.12$.

(b) V is rather small, indicating weak association between the type of gas and the timing of the purchase. There is some association (regular tends to be bought more during the week than premium), but the association is not strong.

43. The expanded table is as follows with the marginal totals:

	Question uses *satisfied*	Question uses *dissatisfied*	Total
Very satisfied	139	128	267
Somewhat satisfied	82	69	151
Somewhat dissatisfied	12	20	32
Very dissatisfied	10	23	33
Total	243	240	483

Chapter 5 Association between Categorical Variables

The table shows the marginal counts.
Combine the first two rows of the table. If the question used the word satisfied, (139+82)/243 = 90.95% were satisfied. If the question used the word dissatisfied, then the percentage satisfied dropped to (128+69)/240 = 82.08%.
Phrase the question in a positive sense using the word satisfied. Wording puts the notion of being satisfied in the customers mind rather than encouraging the customer to think of reasons that he or she might not be satisfied.

45. (a) The table shows weak association. The expected counts from the margins are:

$$(267\cdot 240)/483=134.33 \qquad (267\cdot 240)/483=132.67$$
$$(151\cdot 243)/483=75.97 \qquad (151\cdot 240)/483=75.03$$
$$(32\cdot 243)/483=16.10 \qquad (32\cdot 240)/483=15.90$$
$$(33\cdot 243)/483=16.60 \qquad (33\cdot 240)/483=16.40$$

Accumulating the squared deviations divided by these counts, we obtain

$$\chi^2 = \frac{(139-134.33)^2}{134.33} + \ldots + \frac{(23-16.40)^2}{16.40} = 8.67 \text{ and } V = \sqrt{\frac{8.67}{483\cdot 1}} = 0.13.$$

The association is weak.

(b) The combined table with marginal totals is:

	Question uses *satisfied*	Question uses *dissatisfied*	Total
Satisfied	221	197	418
Dissatisfied	22	43	65
Total	243	240	483

The expected counts are

$$(418\cdot 243)/483=210.30 \qquad (418\cdot 240)/483=207.70$$
$$(65\cdot 243)/483=32.70 \qquad (65\cdot 240)/483=32.30$$

and $\chi^2 = \dfrac{(221-210.30)^2}{210.30} + \ldots + \dfrac{(43-32.30)^2}{32.30} = 8.14$ with $V = \sqrt{\dfrac{8.14}{483\cdot 1}} = 0.13$. Chi-square is slightly smaller, but V is the same.

47. (a) Marginally, given a balanced number of cases in each industry, the marginal proportion of men is 43% ((34%+40%+38%+60%)/4=43%). None of the industries has this proportion, and so each conditional distribution does not match the marginal distribution. There is association.

(b) The table shows association because the proportion of men working in each industry depends on the industry. Some industries have a higher proportion of male employees than female employees.

(c) If $n = 400$ with 100 in each row, then the expected counts are

	Men	Women
Advertising	43	57
Book publishing	43	57
Law firms	43	57
Investment banking	43	57

Accumulating the squared deviations divided by these values, the overall $\chi^2 = \dfrac{(34-43)^2}{43} + \ldots + \dfrac{(40-57)^2}{57} = 16.5$

and Cramer's $V = \sqrt{\dfrac{16.5}{400\cdot 1}} = 0.20$. If $n = 1{,}600$ with 400 in each row, then the expected counts are four times larger. Chi-square is also four times larger, but V is unchanged. Chi-square increases with n, but V does not.

49. (a) The shown proportions differ in the rows of the table, but this is not the type of association that we have studied in this chapter. See part b.

(b) This table is *not* a contingency table. The 2,812 respondents are not put into cells on the basis of two measurements for each. Think of the how the data table is shaped. For each of the 2,812 rows (the survey respondents), we have four columns. One column is the rating assigned to scientists, the second is the rating assigned to banks, and so forth. Each respondent has a value in each row of the table. The cells are not mutually exclusive.

51. (a) The results show clear association, with the appearance that support influences the outcome. Of those that get support, 24 out of 47 (51%) produced a supportive article (percentages within the first row). None of those who were unsupported wrote a favorable paper.
(b) The association is so strong as to make it difficult to find a lurking variable that would mitigate what is shown in the table. That said, one could suggest that the authors were already in the progress of their research prior to funding. Companies only fund results that were headed their way. The support might have come after the research, rather than before.

53. (a) Use the column percentages. United is doing better; its on-time arrival rate overall is $6291/(6291+1466) \approx 0.811$ whereas US Airways is $4,366/(4,366+1,388) \approx 0.759$.
(b) Yes. The on-time percentage at Denver is $5,661/(5,661+1,205) \approx 0.824$ whereas it's $3,649/(3,649+1,284) \approx 0.740$ in Philadelphia. This difference matters because most of the United flights go to Denver, whereas US Airways flies to Philadelphia.
(c) Yes. If we compare the arrival rates at each destination, US Airways comes out better: for example, $5,208/6,323 \approx 82.3\%$ versus $453/543 \approx 83.4\%$ in Denver and $364/510 \approx 71.4\%$ versus $3,285/4,423 \approx 74.3\%$ in Philadelphia.

Chapter 6 Association between Quantitative Variables

Mix and Match

1. (a) ii
   (b) i
   (c) iii
   (d) iv

3. (a) iv
   (b) iii
   (c) i
   (d) ii

True/False

5. True.

7. False, in general. The pattern could also be in a negative direction.

9. True.

11. False. The pattern would be linear, with $y \approx 0.1\,x$. where y denotes revenue and x denotes sales.

13. False. The value of the stock would fall along with the economy. We'd rather have one that was negatively related to the overall economy as a hedge against a recession.

15. True. Untangle the standardized variables and you'll see that the correlation line in the original units predicts y to be $\bar{y} + r\dfrac{s_y}{s_x}(x - \bar{x})$. If $r = 1$, the slope is the ratio of SDs.

17. True. By aggregating to monthly totals, the values of the variables would be about 25 times larger (if the factory operates about 25 days per month). Larger values produce a larger covariance.

19. False. The positive association is more likely the result of rising gasoline prices in the market overall. Over time, prices for gasoline tend to increase. Hence a constant volume of customers produces increasing dollar volume.

Think About It

21. (a) Sales: Total cost is the response and number of items is the explanatory variable. Expect to see a positive direction, linear, with lots of variation because of the varying costs of items bought.
    (b) Productivity: Items produced is the response and hours worked is the explanatory variable. Expect to see positive direction, linear (with perhaps some curvature for long hours), and moderate variation.
    (c) Football: Weight is the explanatory variable and time is the response. Expect negative direction (those little guys need to be quick!), probably linear but with lots of variation in the times.
    (d) Fuel: Number of miles is the explanatory variable and gallons left is the response. Expect negative direction, linear, with small variation around the trend (assuming we drive similarly after each fill-up).
    (e) Investing: The number recommending is the explanatory variable and the subsequent price change is the response. We expect (being skeptical) little or no pattern, but many would expect positive association.

23. (a) There's positive association, but it seems rather weak. It's hard to say whether it's linear from the figure; the relationship is too weak.
    (b) The actual correlation is 0.38.
    (c) The cluster increases the correlation. The correlation without this cluster is about half the size; $r = 0.19$.
    (d) No, when the outliers are excluded, the association is too weak to arrive at this conclusion.

25. The correlation is not affected by changing the scale from either dollars to thousands or dollars to euros.

27. No, because the correlation does not depend on the center of the data. You can add and subtract a constant from a variable without changing the correlation.

29. The slope of the linear relationship measured by the correlation is r. Hence, the predicted z score must be smaller than the observed z score because the absolute value of r is less than 1. This phenomenon is often called regression to the mean.

31. (a) Yes, the two time series appear to move in opposite directions, with inflation gradually rising as consumer sentiment falls. It also appears that peaks in one index coincide with troughs in the other (note particularly the second half of 2005).
    (b) No, the scatterplot clarifies the strength of the association as well as shows a linear pattern.
    (c) The correlation is approximately -0.70.
    (d) The timeplot shows the timing of the events, such as extreme highs, whereas the scatterplot shows the contemporaneous association more clearly and reveals the linear association.
    (e) No. We're still only able to consider association, not causation. Something else (or more likely quite a few other factors) in the economy cause both series to move.

33. The correlation is larger among stones of the same cuts, colors and clarities. These are the other factors that add variation around the correlation line. By forcing these to be the same, we get a more consistent pattern with fewer lurking variables.

35. Cramer's V measures association between *categorical* variables. Because the levels of categorical variables cannot in general be ordered, it does not make sense to speak of the *direction* of the association.

37. You have a 1-in-4 chance of guessing the original. That's not a pattern that you've found; that's luck.

39. (a) Yes, we expect them to be associated, with a moderate correlation on the order of 0.5 or perhaps larger.
    (b) The scatterplot shows very little, if any, association. A couple of stragglers (such as the value for September 19, 2002) float at the edge of the plot, but there's no trend. If anything, the association is negative or perhaps curves like a bowl.

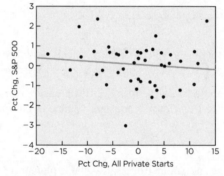

(c) The estimated correlation for these 46 cases is $r = -0.115$, slightly negative.
(d) No, with so little association, it would not be very helpful to predict the movement in the market if you were somehow able to know the housing numbers ahead of time. In slope-intercept form, the slope $b = -0.017$, implying that the associated line (as shown in the scatterplot) is very nearly flat.
$\bar{x} = 0.0713$, $s_x = 6.8835$, $\bar{y} = 0.0137$, $s_y = 1.060$, $b = r \cdot s_y / s_x = -0.115 \cdot 1.060 / 6.8835 \approx -0.017$

41. (a) Yes. Assuming that the employees differ in skill.
    (b) The scatterplot shows strong, positive, linear association. Employees who scored well the first time tend to score relatively well the second time.

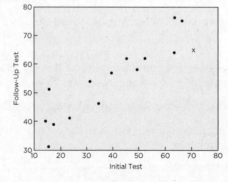

(c) The correlation is $r = 0.903$. The correlation is suitable because the association is linear. We'd expect the relative position on the second test to be lower, closer to 1.8 (twice the correlation) on the second test.

(d) The employee in question is marked by the x in the figure. The correlation line, with slope 0.9 implies that we expect the scores on the second test to be a little closer to the mean than the scores on the first test. The decline for this employee seems consistent with the pattern among the others and it would be inappropriate to judge this employee as becoming less productive.

43. (a) The scatterplot is below, with Entered as the explanatory variable and Errors as the response. We think of errors as the result of entering data rather than the other way around. There's little or no pattern in these data.

(b) 0.094.

(c) The correlation does not depend on the units of the data and would not change.

(d) The correlation indicates a very weak association between the number of data values entered and the number of errors. Evidently, those who enter a lot of values are simply faster and more accurate than those who enter fewer. (They are more accurate because they have entered more items but kept about the same number of errors.)

(e) There's virtually no association between the number entered and the number of errors.

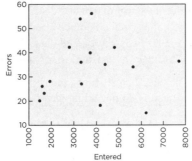

45. (a) The scatterplot of mileage on horsepower (with attached marginal distributions) appears below. This choice for x and y is most natural and works with the setting posed in part e. We generally think of the engine power as explaining the mileage rather than vice versa.

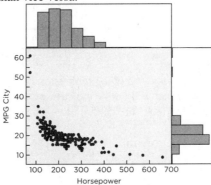

(b) The pattern is negative in direction, bending with a sharp initial drop and then slowing, with relatively little variation around the curving pattern. The outliers on the left with exceptional mileage are hybrids (Honda Insight and Toyota Prius), and those on the right with huge horsepower are exotic sports cars (Lamborghini, Ferrari, etc).

(c) The correlation is -0.69.

(d) The value $r=-0.69$ suggests moderate negative association. It is not a good summary because it does not communicate the strong bending pattern evident in the scatterplot.

(e) Both marginal distributions are right skewed, with extreme outliers (noted in part b). The mean of the horsepower is 214 HP with SD = 84, and the mean of mileage is 19.8 with SD 5.5.

(f) The z score for a car with 125 HP is $z_x = (125 - 214)/84 = -1.06$, a bit more than 1 SD below the mean. We expect its mileage to be about r SDs above the mean (because the correlation is negative). More precisely, $z_x \cdot r = (-1.06) \cdot (-0.69) = 0.73$, giving an estimated mileage of $\bar{y} + 0.73 s_y = 19.8 + 0.73 \cdot 5.5 = 23.8$ MPG. Because the relationship is not linear, this estimate is suspect.

47. (a) The plot on the left below plots the price on the crime rate. The outlier is Center City, Philadelphia, and it is unusual in terms of the crime rate, but not the selling price.

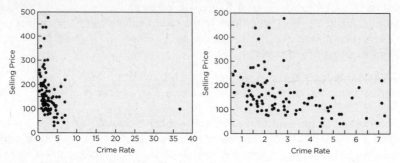

(b) The correlation using all of the data is $r = -0.25$.
(c) The refocused scatterplot on the right shows a great deal of variation around a weak, negative trend that appears to bend. The price seems to drop off faster on the left (few crimes) than the right (more crimes), either that or there are new outliers (such as the cluster of expensive districts at the upper left).
(d) The correlation without Center City, Philadelphia, is much stronger than previously found, -0.43.
(e) No, we cannot for several reasons. First, this is aggregated data. We do not see the prices for individual homes, only for communities. Second, correlation measures association, not causation.

49. (a) The value of chi-square is 8.466. The table below shows the counts, expected count (under no association), and the contribution of that cell to chi-square. Cramer's V is then sqrt(8.466/150) ≈ 0.23757.

Count Expected Cell χ^2	Male	Female	Total
Cash	50 42 1.5238	10 18 3.5556	60
Credit	55 63 1.0159	35 27 2.3704	90
Total	105	45	150

(b) The correlation is approximately 0.24. The following table shows the details.
(c) The two are exactly the same, though that will not be the case when the correlation is negative. Cramer's V is always positive, whereas the correlation picks up the direction. The squared correlation is always the same as the squared value of V.

Variable	Mean	Std. Dev.	Correlation
Sex	0.7	0.459793	0.237566
Cash	0.4	0.491539	

51. This exercise and the next show that macroeconomic time series are often highly correlated with each other because they all measure aspects of a growing economy.
The timeplots, especially if shown together with a separate axis as in the following, are both tending upward, (red for the hourly compensation, and green for the hourly output).

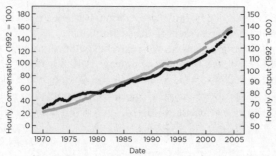

(b) The scatterplot confirms the very high correlation, though the white space rule should make one suspicious.

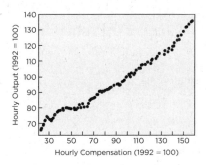

(c) The correlation is 0.988.

(d) Both series track a growing economy that includes the effects of inflation on the stated dollar values. Plus, correlation does not imply causation. It's more the case that compensation and output are both rising as the economy grows and inflation continues.

53. (a) Yes, although one might hope that there would be small correlation, anticipating that the pilots would make up for delays along the way.

(b) The scatterplot shows strong positive linear association, with one pronounced outlier (row 437, a flight on Northwest from Billings, MT to Minneapolis).

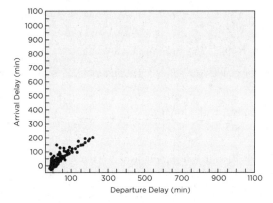

(c) The correlation is 0.958.

(d) The correlation is noticeably smaller without the outlier. It's still rather positive, but has fallen to 0.907.

(e) The correlation would be the same. Correlation has no units, and so it is the same regardless of the time units.

Chapter 7 Probability

Mix and Match

1. j, probabilities multiply for independent events
3. h
5. b
7. d, rearranged to have the intersection on the left hand side.
9. e

True/False

11. False. The sample space consists of all possible sequences of yes and no that he might record. That's $2 \times 2 \times 2 \times 2 \times 2 = 2^5 = 32$ elements in the sample space.

13. False. These are not disjoint events. For example, the outcome that all five shoppers have a bag {yes, yes, yes, yes, yes} lies in both **A** and **B**.

15. True. The probability of the union **A** *or* **B** (cash or $50) is the same or larger than the probability of the intersection **A** *and* **B** (cash and $50).

17. False, both events could happen. Hence, they are not disjoint. If events are disjoint, only one of them can occur.

19. False, only if the data lacks patterns does the relative frequency tend to the probability in the long run.

21. False. The intersection **B** and **C** is a subset of the event **A**. For example, **A** also occurs if 4 of the 6 candidates rating 8 or better come from Monday, with 2 on Tuesday.

Think About It

23. Graphs a and b. There is a clear pattern in c (up then down). There appears to be a slight pattern in d. (gradually flattening, more variable upward trend). There's no pattern in a or b, although there are outliers in b.

25. (a) Intersection: {fresh}
    (b) Union: S = {frozen, refrigerated, fresh, deli}
    (c) Complement: $\mathbf{A}^C$ = {deli}

27. (a) Big (waist) and tall.
    (b) This would mean that the choice of waist size is independent of the length of the pant leg. That is, a tall man is just as likely to be very thin as a short man.
    (c) (**B** *or* **T**). The intersection (**B** *and* **T**) are pants made for large people, those with a thick waist and long legs. The union includes tall people that are skinny as well as short people with a large waist.

29. An intersection. The company wants both attributes (engineering, foreign language), not one or the other.

31. a is true, but not a result of the Law of Large Numbers. b is false unless you have to learn every probability from data. Only c is a consequence of the law of large numbers, and it requires that the trials lack a pattern.

33. Not likely. The intensity of traffic would change over the time of day and the day of the week. Also, the outcomes (whether more than 50 cars pass by) would likely be dependent. Imagine what would happen during a traffic jam or very late at night.

35. (a) Yes. The results appear to lack a disruptive pattern. The clicks are relatively few and far between, but there does not seem to be a pattern in where they occur.
    (b) No. We can only approximate (or estimate) the chance for clicking on the ad from this sequence (because it lacks a visible pattern), but we'd likely get a different answer if we looked at another 100 tosses. We need to see the whole sequence, continuing infinitely long.

37. Go for the 3-point shot. That gives a 30% chance of winning the game. The 2-point strategy gives only a ½ × ½ = ¼ chance of winning the game, assuming that the outcomes are independent so that we can multiply the probabilities.

Chapter 7 Probability

39. Pure speculation, and almost certainly a subjective probability based on this analyst's experience when the market feels like the current situation. Of course, a more quantitative analyst might be using data.

41. (a) The Law of Large Numbers only means that the long run proportion matches the probability in the long run. Additionally, if the accidents are dependent, accidents one day rattle the pilots so much that they do not fly as well the next day, then the LLN does not apply because the process does not consist of independent trials.
(b) No. More likely, accidents form a simple sequence, and one is not safer the day after an accident. You might suspect, however, that after an accident, everyone from mechanics to pilots is more watchful and it might be safer.

You Do It

43. (a) 1. S = {blue, orange, green, yellow, red, brown }
       2. P(blue or red) = P(blue) + P(red) = 0.24 + 0.13 = 0.37
       3. P(not green) = 1 − P(green) = 1 − 0.16 = 0.84
    (b) 1. S = {triples of three colors }
       2. P(blue and blue and blue) = P(blue)^3 = 0.24^3 ≈ 0.0138
       3. P(any color and any color and red) = 1 · 1 · 0.13 = 0.13
       4. P(at least one is blue) = 1 − P(none is blue) = 1 − 0.76^3 ≈ 0.561

45. (a) Assuming the complaints produced by calls are independent and are equally likely to occur for each call to the foreign call center, then
    P(next 3 complain) = P(complain 1st and complain 2nd and complain 3rd)
    = P(complain 1st) · P(complain 2nd) · P(complain 3rd) = 0.62^3 ≈ 0.238
    (b) P(complain 1st and complain 2nd and not complain 3rd) = 0.62^2 · 0.38 ≈ 0.146
    (c) 3 × 0.146 = 0.438 (There are three sequences that produce no complaint: on the first, second or third call.)
    (d) P(none of 10 complain) = 0.38^{10} ≈ 0.0000628

47. (a) If we assume that components fail independently, the probability that the computer works is the probability that all components work, or (999/1000)^{100} ≈ 0.905. This means that the probability of a system failing is about 10%.
    (b) If we want the probability of the system working to be 0.99, then we require that the probability p of a component working (still assuming independence) is p^{100} = 0.99.
    Solving for p by using logs, log p = (log 0.99)/100, we find that p = 0.9999. The rate of defects must be reduced to 1 in 10,000.
    (c) $P(F_1$ or F_2 or ... $F_{100}) \leq 100 \cdot 0.001 = 0.1$. The actual probability of this event found in part a is, assuming independence, slightly smaller at 0.095. The bound is tight in this example because the events have such small individual probability that there's not much double counting.

49. By assuming independence (*i.e.*, not getting rattled), the probability of answering them all correctly is 0.8^6 ≈ 0.262.

51. (a) The probability of staying for more than a year and having a college education is 0.75 × 0.75 = 9/16. This calculation presumes independence, which is questionable here. It could be the case, for example, that those with college education are less likely to stay because of other opportunities.
    (b) This probability is zero, with no qualifications needed. The probability of staying for two years (which is zero) is at least as large as the probability for staying for two year and being college educated.

53. (a) The probability of an Asian message is (76 + 21)/(76 + 21 + 19 + 8) ≈ 0.782
    (b) P(None sent in US or UK) = P(all 3 sent in Asia) ≈ 0.782^3 ≈ 0.478
    (c) There are roughly 58 messages per person in China, so chances are most people are going to get at least one. Pick a person. The chance that they get the first message is 1/(1,306,000,000). The chance they do not get the message is thus close to 1, (1,306,000,000 − 1)/(1,306,000,000). However, if you repeat this 76 billion times the chance of getting missed gets rather small. Your calculator might die if you calculate (1,305,999,999/1,306,000,000)$^{(76 \text{ billion})}$. We were lucky and used logs to work it out to be 5.33 × 10^{-26}. Call it zero if you like.

55. (a) The customer must get 40% or 50% off. These are disjoint events, so the probabilities sum to 1/8.
    (b) The clerk was right to be surprised. We can calculate that the probability of 3 in a row, independently, is

$(1/32)^3 \approx 0.00003$. Not much chance of 3 in a row if the events are truly independent.

(c) Assume that the purchase amount is unrelated to the discount (the card is scratched off at the time of purchase only). Break the overall event into disjoint pieces like this:

P(save more than \$20) = P(50% coupon and buy sweater) + P(more than 10% and buy suit). The only way to save more than \$20 when buying the sweater is to have a 50% coupon. For those spending \$200, the only those with the 10% coupon save \$20 or less. Now use independence:

P(50% coupon and buy sweater) = $1/32 \times \frac{1}{2}$ and P(more than 10% and buy suit) = $\frac{1}{2} \times \frac{1}{2}$.

Hence the overall probability is P(save more than \$20) = $1/64 + 1/4 \approx 0.266$.

57. (a) The plot of the sequence of shots taken does not indicate anything out of the ordinary, especially if you have observed the surprising randomness in the experiments of the previous two questions.

(b) One explanation for the cooling down at the end involves fouls by the opposing team. Bryant scored 12 points on free throws in the 4th period alone (18 points for the game). Misses associated with fouls are counted in the box score as misses just the same. Bryant was fouled on the last 3 shots (which were misses).

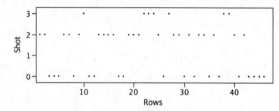

Chapter 8 Conditional Probability

Mix and Match

1. f

3. c

5. b

True/False

7. False. The statement asserts that $P(A) > P(A|S)$. This might happen, but need not be the case.

9. False. The statement asserts that $P(A) > P(S)$, but these are only marginal probabilities and the ordering does not imply dependence.

11. False. She also needs a joint probability.

13. False. Independence implies that one event does not influence the chances for the other.

15. False. These are disjoint events, so the probability of the intersection must be 0.

17. True. Think about how the mosaic plot looks when there is independence.

19. True.

Think About It

21. Independent. It is unlikely that seeing a Honda, for example, would make us suspect that the next car was also a Honda (unless there's a parade of Hondas which is unlikely on an interstate highway).

23. Dependent. The number of visits today is probably influenced by the same factors that influenced the number of visits yesterday.

25. Most likely independent, unless we know that there is some particular sale in progress that has drawn shoppers looking for the same item.

27. Independent. If **A** happens, then the chance for **B**, $P(B|A)$, remains 1/4 because **B** is ¼ of the area of **A**.

29. (a) Classify everyone as a drug user. In this way the sensitivity of the test must be 1, P(test says drug use | use drugs) = 1.
    (b) The problem is that the test will not be very specific; the test will have many false positives. In particular, any clean person who takes the test will be falsely accused of using drugs.

31. No. The statement of the question first gives $P(F|Y) = 0.45$ and then $P(Y|F) = 0.45$. In order to be independent, $P(F|Y) = P(F)$. We are not given marginal probabilities and hence cannot determine that the events are independent. To see some pictures, have a look at the Venn diagrams for Exercises 27 and 28. In both cases, $P(A|B) = P(B|A)$, but in Exercise 28 the events are independent whereas in Exercise 27 they are dependent.

33. (a) 0.42 = P(working affected grades | have loan). The sample space might be the collection of recent college grads.
    (b) You cannot tell. In general, you cannot obtain $P(B|A)$ from $P(A|B)$ without the marginal probabilities. In this example, you don't know the proportion who worked in college.

You Do It

35. (a) The choice is up to you, but a tree is easy to use in this context for two main reasons: the probabilities are given in a sequential form (as conditional probabilities), and the options for the two types of vehicles are *not* the same. You cannot get a sunroof on a truck from this brand, but you can get one on a car.
    (b) P(sunroof) = P(car) × P(sunroof | car) = 1/2 × 1/4 = 1/8.

37. P(drink *and* popcorn) = P(drink) P(popcorn | drink) = $0.7 \times 0.3 = 0.21$

39. You need to reverse the conditioning. This problem can be solved by Bayes' Rule. A table organizes the information easily without the need to remember the formula and provides a check on the calculations. If a player fails (the first column), then there's a 0.50/0.55 ≈ 0.91 chance that it has the flaw.

39. You need to reverse the conditioning. This problem can be solved by Bayes' Rule. A table organizes the information easily without the need to remember the formula and provides a check on the calculations. If a player fails (the first column), then there's a $0.50/0.55 \approx 0.91$ chance that it has the flaw.

	Fails in Six Months	Does Not Fail	Total
Has flaw	0.50	0	0.50
No flaw	0.05	0.45	0.50
Total	0.55	0.45	1

41. We assume that any order of selection of the parts is equally likely.
    (a) P(both good) = P(first good) × P(second good | first good) = (7/12) (6/11) = 7/22
    (b) P(at least one is good) = 1 − P(none is good)
    $= 1 - P$(first bad) P(second bad | first bad)P(third bad | first two bad)|
    $= 1 - (5/12)(4/11)(3/10) = 21/22$
    (c) P(four good) = (7/12)(6/11)(5/10)(4/9) = 7/99
    (d) P(four bad then good) = (5/12)(4/11)(3/10)(2/9)(7/8) = 7/792 ≈ 0.00884

43. (a) 1/12
    (b) 1/11
    (c) The events are dependent. Let the event **A** denote finding the system with the missing component first, and **B** denote finding it second. Then $P(\mathbf{B}|\mathbf{A})$ is 0 because he cannot find it on the second system if he already found it on the first. In lay language, he has 12 possible choices first, but only 11 possible choices second. He's more likely to find it second since he's eliminated one of the extra choices.

45. (a) Dependent. The probability that the first assignment goes to a man is 25/50, but this event changes the probability for the second assignment (because there is one less man who can receive the assignment). The probability that the second assignment also goes to a man given that the first is to a man is 24/49 < 1/2.
    (b) P(fifth lead to a man | first four leads go to men) = 21/46
    (c) You can do this one several ways. Let **A** denote the event that at least four leads go to men, and let **B** denote the event that men receive all five leads. The question asks for $P(\mathbf{B}|\mathbf{A}) = P(\mathbf{B}$ and $\mathbf{A})/P(\mathbf{A}) = P(\mathbf{B})/P(\mathbf{A})$ (**B** is a subset of **A**). Let's find $P(\mathbf{A})$ first. If W stands for a lead given to a woman and M for a lead given to a man, then the the event **A** means that the final result must be one of these.
    WMMMM, MWMMM, MMWMM, MMMWM, MMMMW, MMMMM
    The chances for any of the events with a woman getting one lead are the same. For example, multiplying the conditional probabilities, P(MMWMM) = (25/50)(24/49)(25/48)(23/47)(22/46) ≈ 0.02985
    and P(MMMMW) = (25/50)(24/49)(23/48)(22/47)(25/46) ≈ 0.02985.
    Now that you have seen two, you can see why all five are the same. No matter where we put the W, the probability is always
    $$\frac{25 \times 24 \times 23 \times 22 \times 25}{50 \times 49 \times 48 \times 47 \times 46}$$
    The probability for the last outcome is
    $P(\mathbf{B}) = P$(MMMMM) = (25/50)(24/49)(23/48)(22/47)(21/46) ≈ 0.02508
    Hence the probability of four or more men is (the events are disjoint, and so add)
    $5 \times 0.02985 + 0.02508 = 0.17433$
    The probability of five men given at least four men is then
    $P(\mathbf{B})/P(\mathbf{A}) = 0.02508/0.17433 \approx 0.1439$

Chapter 8 Conditional Probability

47. The following table converts the given conditional probabilities into counts (in millions) for each occupation. From these counts, we can define the joint distribution associated with randomly picking a worker from this collection.

	1965	2005	Total
Durable goods manufacturing	13.3	11.2	24.5
Professional services	5.6	21.0	26.6
Education and health care	4.9	22.4	27.3
Other occupations	46.2	85.4	131.6
Total in workforce	70.0	140.0	210.0

$P(1965|\text{durable}) = 13.3/24.5 \approx 0.543$
(b) $P(1965|\text{services, education, or health}) = (5.6+4.9)/(26.6+27.3) \approx 0.195$
(c) If they say education/health care, you're most likely in 2005. The conditional probabilities of the year are most different for this occupation. $P(1965 | \text{Educ, health}) = 4.9/27.3 \approx 0.179$ compared to

49. A table makes it easy to organize the information. The values in the following table in bold come directly from the statement of the exercise. The rest come from making the values add up correctly.
    (a) $P(\text{Luggage in SF}) = 0.82$
    (b) $P(\text{Arrived late in Dallas}|\text{No luggage in SF}) = 0.1/0.18 = 5/9$

	Arrive on Time	Arrive Late	Total
Luggage	0.72	0.1	0.82
No luggage	0.08	0.1	0.18
Ttal	**0.8**	0.2	1

51. It helps in problems such as this to organize the information in a table. We are given the information shown in bold in the table below. We are also given $P(\text{No internet} | \text{Computer}) = 0.18$. The other probabilities are derived as follows:

$P(\text{No internet}) = 1 - P(\text{Internet}) = 0.25$
$P(\text{Computer}) = P(\text{Computer and Internet}) + P(\text{Computer and No internet})$
$= 0.75 + P(\text{No internet} | \text{Computer}) \cdot P(\text{Computer})$
$= 0.75 + 0.18 P(\text{Computer})$

Thus, $P(\text{Computer}) \approx 0.91$.

	Internet Access	No Internet	Total
Have computer	0.75	0.165	0.915
No computer	0	0.085	0.085
Total	**0.75**	0.25	1

It then follows that $P(\text{No computer} | \text{No internet}) = 0.085/0.25 = 0.34$.

Chapter 9 Random Variables

Mix and Match

1. g

3. f

5. b

7. d

True/False

9. True.

11. False. The mean of X should be smaller than the mean of Y.

13. False. The mean is the weighted average of possible outcomes with all weights between 0 and 1; it need not be one of the outcomes.

15. False. Properties of the mean of a random variable are comparable to means of the data. If, for example, the probability distribution is skewed to the left, then more than half of the probability (as with histograms of dat(a) is to the left of the mean.

17. True.

19. True, assuming all other things remain the same.

Think About It

21. $P(X = -2) = 0.3$.

23. $P(Z \leq -3) + P(Z = 3) = 0.05 + 0.05 = 0.10$.

25. $E(Z) > 0$ since the probabilities of the positive values are larger than the corresponding negative probabilities.

27. Y has the largest SD. The uniform distribution spreads the values farther from the center than the others that have some clustering.

29. 0. Notice that $P(Y \leq 0) = ½$.

31. (a) $P(W=5) = p(5) = 1/10$ and $P(W = -1) = p(-1) = 9/10$. The probability distribution has two non-zero values, with the height at $w = 5$ being 9 times higher than the other.
    (b) The expected value of W is $5 \times (1/10) + (-1)(9/10) = -4/10 = -0.4$. The game is not fair because the mean is not zero.

33. (a) Fair. The player has half of the probability and contributes half of the pot.
    (b) Better than fair to the player. The player contributes 1/100 of the pot but has a larger share of the probability of winning (1/52).

You Do It

35. The means and standard deviations are

	μ	σ
X/3	40	5
2X − 100	140	30
X + 2	122	15
X − X	0	0

37. (a) There are three possible values for the investor, depending on whether both stocks rise, both fall or one goes up while the other goes down.

Chapter 9 Random Variables

Outcome	P(X)	X
Both increase 80% to $18,000	1/4	$36,000
One increases	1/2	$22,000
Both fall 60% to $4,000	1/4	$8,000

(b) E(X) = 36,000/4 + 22,000/2 + 8,000/4 = $22,000.
(c) Yes, the probability is symmetric around the mean gain of $2,000.

39. Let the random variable X denote the earned profits. Then the probability distribution of X is $p(0) = P(X = 0) = 0.05$, $p(20,000) = 0.75$, $p(50,000) = 0.20$.
(b) The expected value of X is $(0)(0.05) + (20,000)(0.75) + (50,000)(0.20) = \$25,000$.
(c) The variance of X is $(0-25,000)^2 (0.05) + (20,000-25,000)^2 (0.75) + (50,000-25,000)^2 (0.20) = 175,000$.
Hence $\sigma = \sqrt{175,000} \approx \$13,229$.

41. (a) $E(X) = 0 (0.05) + 1 (0.25) + \ldots + 5 (0.05) = 2.25$ reams.
(b) $Var(X) = (0-2.25)^2 (0.05) + (1-2.25)^2 (0.25) + \ldots + (5-2.25)^2 = 1.5875$ so that
$$\sigma \approx 1.26 \text{ reams.}$$
(c) $E(20 - X) = 20 - 2.25 = 17.75$ reams.
(d) 1.26 reams (same as part b because adding or subtracting a constant has no effect).
(e) $E(500X) = 500E(X) = 1,125$ pages; $SD(500X) = 500SD(X) \approx 630$ pages.

43. (a) Let R be a random variable that denotes the bolivar/dollar exchange rate in six months. The problem indicates that $P(R=2.15) = 0.6$ and $P(R=5) = 0.4$. The expected current value of the contract in dollars is then (divide by the rate to get the cost in dollars)
$$E(1,000,000/R) = 1,000,000 \, E(1/R) = 1,000,000 \, ((1/2.15) \times 0.6 + (1/5) \times 0.4)$$
$$\approx 1,000,000 \, (0.359) = \$359,000.$$

(b) Dividing the cost in bolivars by the expected value of the exchange rate gives $1,000,000/3.29 \approx \$304,000$. This is substantially less than the expected cost calculated correctly in part a. The error is that $E(1/R) > 1/E(R)$.

45. (a) Let $X = 0, 1$ denote whether a customer buys the printer when no rebate is offered (1 for yes and 0 for no). The expected value of X is $E(X) = 0 \times (1-p) + 1 \times p = p$. Similarly, let X^* denote whether a customer who is offered a rebate buys a printer; $E(X^*) = p^*$.
If the company does not offer a rebate, its profits are $60X$ with expected value
$$E(60X) = 60p.$$
If the company offers the rebate, the expected profits are
$$E(30X^*) = 30p^*.$$
For the rebate to be effective, on average, $30 p^* > 60 p$, so $p^*/p > 60/30 = 2$; the probability of purchase has to double.
(b) Let $Y = 0, 1$ denote whether the rebate is used. Then the expected profit when the rebate is offered (assuming X^* and Y are independent) is a random variable Z with probability distribution
$$P(Z = 0) \quad = 1 - p^*$$
$$P(Z = 30) \quad = 0.4p^* \text{ (purchase and use rebate)}$$
$$P(Z = 60) \quad = 0.6p^* \text{ (purchase and do not use rebate)}$$
The expected profits are then $E(Z) = 30 \times 0.4p^* + 60 \times 0.6p^* = 48p^*$. For the rebate to be effective, $48p^* > 60p$ so $p^*/p > 60/48 = 1.25$, quite a bit smaller than in part a.

47. (a) Let X denote the number of clients visited each day. Then the only way that he sees one client is if that client buys a policy. Hence, $P(X = 1) = 0.1$. He sees two clients if the first does not buy a policy, but the second does. Otherwise, he sees three clients.
$$P(X = 1) = 0.1$$
$$P(X = 2) = 0.09$$
$$P(X = 3) = 1 - p(1) - p(2) = 0.81.$$
(b) $E(X) = 1 \times 0.1 + 2 \times 0.09 + 3 \times 0.81 = 2.71$ clients.

(c) $2.5 \times E(X) = 6.775$ hours.
(d) He either sells nothing or one policy each day (since he stops after the first sal(e). Hence, his expected earnings are $3000 times the probability of a sale, which can be found as the sum of the probabilities of three disjoint events:

$P(sal(e)) = P(\text{first buys}) + P(\text{second buys}) + P(\text{third buys}) = 0.1 + 0.09 + 0.081$
$= 0.271$.

Hence, his expected earnings per day are $3000 \times 0.271 = \$813$.

49. (a) The probability distribution of the ATM withdrawal is

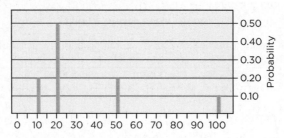

(b) $p(50)+p(100) = 0.3$.
(c) $E(X) = 10 \times 0.2 + 20 \times 0.5 + 50 \times 0.2 + 100 \times 0.1 = \$32$.
(d) On average, we expect a customer to withdraw $32. This is the average of the outcomes in the long run, not a value that occurs for any one customer.
(e) $E(X-32)^2 = (10-32)^2\, 0.2 + (20-32)^2\, 0.5 + (50-32)^2\, 0.2 + (100-32)^2\, 0.1 = 696$
so the SD of X is $\sqrt{696} = \$26.4$.

51. (a) Two free throw attempts have the largest expected value:
   $E(\text{two free throw attempts}) = 2 \times 382/481 \approx 1.59$ points
   $E(\text{one field goal attempt}) = 2 \times 462/990 \approx 0.93$ points
   $E(\text{one 3-pt attempt}) = 3 \times 47/133 \approx 1.06$ points
(b) As seen in part a, both of these shots have similar expected values. The variances, however are rather different with much more variation associated with the 3-point attempt.
   $\text{Var}(\text{2-pt attempt}) = (0-0.93)^2 \times (528/990) + (2-0.93)^2 \times (462/990) \approx 0.996$
   $\text{Var}(\text{3-pt attempt}) = (0-1.06)^2 \times (86/133) + (3-1.06)^2 \times (47/133) \approx 2.06$
Though similar in expected value, 3-pointers bring more variation and perhaps excitement.
(c) Two free throws. The calculation for free throws now does require independence as well as equal chances for making the shot.
   $P(\text{2 points by free throws}) = $ (if in(d) $= (382/481)^2 \approx 0.63$
   $P(\text{2 points by field goal}) = 462/990 \approx 0.47$

53. (a) This histogram is rather bell shaped and centered nearly at zero, with almost all of the daily returns within 5% of zero.

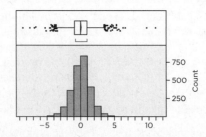

(b) The mean is $\bar{x} = 0.0652$ with $s = 1.49$.
(c) Given the rate 0.02 for the loan, the Sharpe ratio is $(0.0652-0.02)/1.49 \approx 0.030$
(d) 0.030. The Sharpe ratio is the same since it uses the percentage changes and does not depend on the amount that is invested.
(e) We first have to compute the Sharpe ratio for IBM using the same bank interest rate. That change implies that the Sharpe ratio for IBM is $(0.10-0.02)/2.24 = 0.036$. The Sharpe ratio for Exxon is lower, suggesting to an investor that IBM offers a better return for the risks.

(f) That performance of these stocks in the future will resemble the past values.
(g) The time plot shows that the variation in the returns was higher in the 1990s. The periods of higher variation (or higher volatility) are clustered; a pattern, but in the variation rather than the mean.

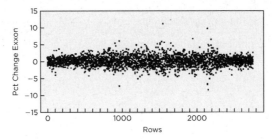

Chapter 10 Association between Random Variables

Mix and Match

1. e. Positive dependence increases the variance of a sum.
3. h. The covariance is zero if the random variables are uncorrelated; hence, the variance of the sum is the sum of the variances.
5. i
7. f
9. a

True/False

11. False. If the costs move simultaneously, they should be treated as dependent random variables.
13. False. The mean and variance match, but this is not enough to imply that all of the probabilities match and that $p(x) = p(y)$.
15. False. This implies that X and Y have no covariance, but they need not be independent.
17. True. $E(X_1) = E(X_2) = \mu$.
19. False. The SD of the total is $\sqrt{2}$ times σ. The variance of the total sales is 2 times σ^2.
21. False. If the effect were simply to introduce dependence (but not otherwise alter the means and SDs), then the expected difference would remain zero.

Think About It

23. Negative covariance produces smaller variance for the portfolio, and hence less risk.
25. The covariance between Y and itself is its variance. The covariance between X and Y is the expected value $E(X-\mu_x)(Y-\mu_y)$. If the two random variables are the same, then $E(Y-\mu_y)(Y-\mu_y) = E(Y-\mu_y)^2 = Var(Y)$. The correlation is 1.
27. No. The covariance depends on the units used to measure the investments. If we track the prices in pennies, say, rather than dollars, we can make the covariance get very large.
29. Not likely. Sales during the weekend would probably look rather different than those during the week.
31. (a) Positively correlated. It seems unlikely that a homeowner would spend a lot for labor but scrimp for the appliances. An elaborate project would generate higher costs in both categories, unless there are budget constraints.
    (b) A budget constraint may produce negative association or independence. If the family spends a lot on appliances, for example, then they will not have much left over for labor.

You Do It

33. (a) $E(2X - 100) = 2E(X) - 100 = 1,900$
       $Var(2X - 100) = 4Var(X) = 4(200)^2 = 160,000$; $SD(2X - 100) = 400$
    (b) $E(0.5Y) = 0.5E(Y) = 1,000$
       $Var(0.5Y) = 0.25Var(Y) = 0.25(600)^2 = 90,000$; $SD(0.5Y) = 300$
    (c) $E(X + Y) = E(X) + E(Y) = 3,000$
       $Var(X + Y) = Var(X) + Var(Y) = 200^2 + 600^2 = 400,000$; $SD(X + Y) \approx 632.5$
    (d) $E(X - Y) = E(X) - E(Y) = -1,000$
       $Var(X - Y) = Var(X) + Var(Y) = 200^2 + 600^2 = 400,000$; $SD(X + Y) \approx 632.5$

35. All of the calculated expected values remain the same. Only the variances and standard deviations change. Unless both X and Y appear, the variance is the same.
    (a) Unchanged
    (b) Unchanged

Chapter 10 Association between Random Variables

(c) $\text{Var}(X + Y) = \text{Var}(X) + \text{Var}(Y) + 2\text{Cov}(X, Y) = 200^2 + 600^2 + 2 \times 12{,}500 = 425{,}000$; $\text{SD}(X + Y) \approx 651.9$

(d) $\text{Var}(X - Y) = \text{Var}(X) + \text{Var}(Y) - 2\text{Cov}(X, Y) = 200^2 + 600^2 - 2 \times 12{,}500 = 375{,}000$; $\text{SD}(X + Y) \approx 612.4$

37. From the formula for the variance of a sum, the covariance is half of the difference, $\text{Cov}(X, Y) = \frac{1}{2}[\text{Var}(X + Y) - \text{Var}(X) - \text{Var}(Y)] = \frac{1}{2}(8 - 10) = -1$. The correlation is $\text{Cov}(X, Y)/[\text{SD}(X)\text{SD}(Y)] = -1/5 = -0.2$.

39. (a) These are dependent, because for example we can write $Y = 60 - X$. Once we know X, Y is known as well from this formula.
    (b) The covariance lowers the variance. After all, $X + Y = 60$ and so the variance is 0.

41. (a) Let X_1 and X_2 denote the deliveries for the two. Both drivers are said to operate independently. We assume, given nothing else, that the number of deliveries is comparable as well.
    (b) $E(X_1 + X_2) = 6 + 6 = 12$ deliveries
    $\text{SD}(X_1 + X_2) = \sqrt{2^2 + 2^2} = \sqrt{8} \approx 2.83$ deliveries
    (c) $E(X_1 + 1.5X_2) = 6 + 1.5 \times 6 = 15$ hours
    (d) $\text{SD}(X_1 + 1.5X_2) = \sqrt{2^2 + (1.5^2)(2^2)} = \sqrt{13} \approx 3.61$ hours
    (e) Yes. It suggests that the counts of the number of deliveries are negatively correlated and not independent, as if the two drivers split a fixed number of deliveries each day.

43. (a) $E(X) = 1(0.50) + 2(0.50) = 1.5$ sandwiches
    $\text{Var}(X) = (1 - 1.5)^2(0.50) + (2 - 1.5)^2(0.5) = 0.25$ sandwiches^2
    (b) $E(Y) = 1(0.60) + 2(0.35) + 3(0.05) = 1.45$ drinks
    $\text{Var}(Y) = (1 - 1.45)^2 (0.60) + (2 - 1.45)^2 (0.35) + (3 - 1.45)^2 (0.05) = 0.3475$ drinks^2
    (c) $E(XY) = 1(0.4) + 2(0.2 + 0.1) + 4(0.25) + 6(0.05) = 2.3$
    $\text{Cov}(X, Y) = 2.3 - 1.5(1.45) = 0.125$ sandwich-drinks
    $\text{Corr}(X, Y) = \text{Cov}(X, Y)/(\sigma_x\sigma_y) = 0.125/(0.5 \times 0.589) \approx 0.424$
    (d) Customers who buy more drinks also buy more sandwiches. The two are positively related, so those who buy more of one also purchase more of the other.
    (e) $E(1.5X + 1Y) = 1.5(1.5) + 1(1.45) = \$3.70$
    $\text{Var}(1.5X + 1Y) = 1.5^2\text{Var}(X) + \text{Var}(Y) + 2(1.5)(1)\text{Cov}(X, Y) = 2.25(0.25) + 0.3475 + 3(0.125) = 1.285\ \$^2$
    Hence, $\text{SD}(1.5X + 1Y) \approx \$1.13$.
    (f) $E(Y/X) = 1(0.40 + 0.25) + 2(0.1) + 0.5(0.2) + 1.5(0.05) = 1.025$
    The ratio of means, $\mu_y/\mu_x = 1.45/1.5 \approx 0.97$, is less than 1. These do not agree. In general, for positive random variables, $E(Y/X) \geq E(Y)/E(X)$.

45. (a) From the numbers given for the 2004 to 2005 season, we get
    $E(X) = 0(689/1376) + 2(687/1376) \approx 0.999$
    $\text{Var}(X) = (0 - 0.999)^2 (689/1376) + (2 - 0.999)^2(687/1376) \approx 1.000$
    (b) $E(Y) = 0(200/308) + 3(108/308) \approx 1.052$
    $\text{Var}(Y) = (0 - 1.052)^2(200/308) + (3 - 1.052)^2(108/308) \approx 2.049$
    (c) Let $X_1, X_2 \ldots X_{20}$ denote the 20 attempted two-point shots and let $Y_1, Y_2, \ldots ,Y_5$ denote the three-point attempts. Then the total number of points scored is
    $T = X_1 + X_2 + \ldots + X_{20} + Y_1 + \ldots + Y_5$
    Assuming the X's are identically distributed and the Y's are identically distributed,
    $E(T) = 20\mu_x + 5\mu_y \approx 20(0.999) + 5(1.052) = 25.24$
    (d) To compare, we need the standard deviation of the points scored from 2 and 3-point baskets. If we assume independence of the shots, then
    $\text{Var}(T) = 20\text{Var}(X_1) + 5\text{Var}(Y_1) \approx 20(1.000) + 5(2.049) = 30.245$
    where we get the variances of these from the calculations in parts a and b.
    Hence, we can see that the $\text{SD}(T) \approx 5.5$. His scoring in this game is within 1 SD of his season performance, so it seems typical from what we are given.

47. (a) No. Larger homes with more occupants would typically use more of both. Expect positive dependence.
    (b) The total cost is $T = 0.09X + 10Y$. Hence the expected total cost is
    $E(0.09X + 10Y) = 0.09(12{,}000) + 10(85) = \$1{,}930$
    (c) The covariance between X and Y is $\rho\sigma_x\sigma_y = 0.35(2{,}000)(15) = 10{,}500$ KWH–MCF
    Using this, we can find that
    $\text{Var}(T) = 0.09^2\text{Var}(X) + 10^2\text{Var}(Y) + 2(0.09)(10)\text{Cov}(X, Y)$

$$= 0.09^2(2{,}000)^2 + 10^2(15)^2 + 2(0.09)(10)10{,}500 = 73{,}800 \text{ \$}^2$$

so that SD(T) ≈ \$272.

(d) The costs for these homes are only about $(1{,}930 - 1{,}640)/272 = 1.1$ SDs above the national figure for 2001. That was a long time ago, so these numbers seem atypical.

49. Let X denote the number of weeks of electrical work, and Y denote the number of weeks of plumbing work.

    (a) $E(X + Y) = 64 + 120 = 184$ weeks

    (b) Positive, because delays or extra time for one type of work suggest similar problems in the other type of work as well.

    (c) $\text{Var}(X + Y) = 6^2 + 15^2 + 2\rho(6)(15) = 387$ weeks^2 and SD($X + Y$) ≈ 19.7 weeks.
    The covariance between X and Y is $\rho\sigma_x\sigma_y = 63$.

    (d) Less. The larger ρ, the more variable the amount of work, making the bidding process less accurate and harder for the firm to estimate its profits.

    (e) $E(200X + 300Y) = 200(64) + 300(120) = \$48{,}800$
    $$\begin{aligned}\text{Var}(200X + 300Y) &= 200^2\text{Var}(X) + 300^2\text{Var}(Y) + 2(200)(300)\text{Cov}(X, Y)\\ &= 40{,}000(36) + 90{,}000(225) + 120{,}000(63) = 29{,}250{,}000 \text{ \$}^2\end{aligned}$$
    SD($200X + 300Y$) ≈ \$5,408.

    (f) The firm is unlikely to make \$60,000 because this amount is more than 2 SDs above the mean.

Chapter 11 Probability Models for Counts

Mix and Match

1. i

3. j

5. c

7. a

9. f

True/False

11. False. These 25 make up a small portion of the total number of transactions (this application of the binomial does not violate the 10% condition).

13. False. The chance that the first three are OK is $(999/1{,}000)^3$. Hence, the chance for an error is $1 - (999/1{,}000)^3 \approx 0.003$. Boole's inequality (Chapter 7) works nicely here.

15. False. This clustering would likely introduce dependence (such as several related transactions that were entered together incorrectly).

17. False. We typically need to work to assure independent trials. A Poisson model requires independent events, not clusters.

19. True. The rate would increase from $\lambda = 0.24$ for 2 square yards to 0.48 for 4 square yards.

21. False. The rate $\lambda = 0.24$. Hence the chance for no defect is $\exp(-0.24) \approx 0.787$. The chance for at least one defect is 0.213.

Think About It

23. (a) Yes, unless we have reason to believe that the cars are traveling together. Each car represents a new Bernoulli trial (fill-up or not).
(b) No. There is little reason to think that the reactions of the directors are independent of one another or have an equal chance of being in favor of the proposal.
(c) Yes, unless there is some hidden dependence in the sealing mechanism that might cause it to stop working altogether and fail to package an entire case.

25. Both are equally likely under the assumption of Bernoulli trials, each with probability $p^3(1-p)^3$.

27. These are equally likely, both with probability $(\tfrac{1}{2})^4$.

29. No, it's quite a bit smaller, $1/2^{10} \approx 0.001$.

31. $3 \times 6 = 18$

You Do It

33. $X \sim \text{Bi}(25, 0.10)$
(a) Yes. The dependence is not a problem because the sampled transactions represent such a small portion of the total collection. There is some dependence, but it is small.
(b) We expect the auditor to find $(0.10)(25) = 2.5$ such transactions. The binomial model concentrates near its mean, so we would expect more than a 50% chance for more than two.
(c) $P(X = 0, 1, \text{ or } 2) \approx 0.0718 + 0.1994 + 0.2659 \approx 0.537$ so that the probability of more than 2 is $1 - 0.537 = 0.463$.

35. (a) A binomial or Poisson model, with Poisson a natural choice as an approximation because n is large and p is small.
(b) Using the Poisson model, $X \sim \text{Poi}(\lambda = 75 \times 0.001 = 0.075)$, $P(X \leq 1) = \exp(-\lambda)(1 + \lambda) \approx 0.9973$. Hence, $P(X \geq 2) \approx 0.0027$.

37. The cost of the first policy is $1,000 plus $1,000 times the number of accidents (call this X), so it has expected costs
$$E(1{,}000 + 1{,}000X) = 1{,}000 + 1{,}000\mu$$
Similarly, the cost for the second policy is
$$E(3{,}000 + 250X) = 3{,}000 + 250\mu$$
The break-even point is $\mu = 2000/750 \approx 2.67$ accidents per year. Choose the $3,000 up-front policy (the second one) if you expect to have more than 2.67 accidents per year.

39. Let $X \sim \text{Bi}(12, 0.10)$ and $Y \sim \text{Poisson}(\lambda = 1.2)$.
    (a) $P(X = 0) = 0.9^{12} \approx 0.282$
    (b) $P(Y = 0) = \exp(-1.2) = 0.301$.

41. Let $X \sim \text{Bi}(20, 0.35)$
    (a) We need to assume that the shots are all made with 35% accuracy and independently (no hot streaks). It's a popular model, but some baskets shot from farther away are harder than those made close to the basket (such as a layup).
    (b) $E(X) = 0.35 \times 20 = 7$
    (c) Yes. Rather than adding the probabilities, consider the empirical rule. $SD(X) = \sqrt{20(0.35)(0.65)} \approx 2.1$. Making 12 or more is 2 SDs above the mean, so it's not very likely. The probability is 0.0196 (from the binomial distribution).
    (d) $E(2X) = 14$
    (e) Let $Y = 2$ or 3, with equal probability and independently of X. Then the expected number of points scored is $E(YX) = E(Y)E(X) = 2.5 \times 7 = 17.5$.

Chapter 12 The Normal Probability Model

Mix and Match

1. f

3. a

5. h

7. e. A googol is 10^{100}, and $P(Z < -20)$ is approximately 10^{-90}.

9. j

True/False

11. False. The probability of the region between 38 and 44 corresponds to $P(0 < Z < 1) \approx 1/3$ whereas the region above 44 corresponds to $P(Z > 1) \approx 1/6$.

13. True. Converting to days amounts to multiplying X by 365. The mean and SD change, but the distribution remains normal.

15. True. If $X \sim N(38, 6)$, then $P(X < 30) \approx 0.0912$. Times 400 gives the expected count.

17. True. As noted in the discussion of the Central Limit Theorem, sums of normal random variables are normal.

19. False. The statement is true on average, but not specifically for tomorrow and the day after. Both random variables have the same distribution, but that does not mean that they are the same. These are like two tosses of the same coin: the chances are the same, but the outcomes need not match.

21. True. $P(X_2 > X_1) = P(X_2 - X_1 = 0) = \frac{1}{2}$ since the difference is normal with mean zero.

Think About It

23. The normal distribution puts no limits on the size of the possible values. Even a standard normal with mean 0 and variance 1 can be arbitrarily large, though with tiny probability.

25. These data are skewed to the right, so we can guess that $K_3 > 0$. In fact, $K_3 = 1.1$ for these data.

27. Both are normal with mean 2μ, but the variance of the sum is $2\sigma^2$ rather than $4\sigma^2$.
    $E(2X_1) = 2E(X_1) = 2\mu$ and $E(X_1 + X_2) = E(X_1) + E(X_2) = \mu + \mu$, whereas
    $Var(2X_1) = 4Var(X_1) = 4\sigma^2$ and $Var(X_1 + X_2) = Var(X_1) + Var(X_2) = 2\sigma^2$
    Think of $2X_1 = X_1 + X_1$ as the sum of two perfectly dependent normal random variables whereas $X_1 + X_1$ is the sum of two independent normal random variables.

29. A Skewed; B Outliers; C Normal; D Bimodal

31. (a) A is the original data and B is with rounding. The rounding shows up as small gaps, stairsteps in the quantile plot.
    (b) The histograms are so similar because the amount of rounding is small relative to the size of the bins.

33. (a) Yes, so long as the weather was fairly consistent during the time period. If, on the other hand, a strong heat wave caused temperatures to soar, then these would create the sort of outliers that the normal model would not accommodate.
    (b) No, this would probably not be appropriate because the weather is a dependent process. The use on a given day is most likely highly dependent on the use on the previous and next days. The amounts might be normally distributed, but not independent.

35. (a) $300 is 1 SD below the mean, so we'd expect about 1/6 to earn less than $300, or about 5/6 to earn more. (If you used a table, you'd get 84.1%.)
    (b) It shifts the mean by 100, from $700 to $800.
    (c) The mean increases by 5% from $700 to $735. The SD also increases by 5%, from $400 to $420.
    (d) No. It appears that the distribution is skewed to the right (which is what you would expect for the

distribution of salaries). The mean is substantially larger than the median and there's relatively little data to the left of the mean (much less than indicated by the calculation in part (a)).

37. (a) Because the homes in the development are generally rather similar but for minor differences, a normal model is reasonable. The price of each is the overall average plus various factors that increase and decrease the value of each.
(b) Sales data from recent housing projects recently built by this contractor with similar characteristics. If those are not available, then sales data from homes with similar types of construction in the area.
(c) Normal with mean $400,000 and SD $50,000. The range $\mu \pm \sigma$ = [$400,000 to $500,000] would then be expected to hold 2/3 of the distribution of prices.

You Do It

39. (a) $P(Z < 1.5) = 0.93319$
(b) $P(Z > -1) = 0.8413$
(c) $P(|Z| < 1.2) = 0.7699$
(d) $P(|Z| > 0.5) = 0.6171$
(e) $P(-1 \leq Z \leq 1.5) = 0.93319 - 0.1587 = 0.77449$

41. (a) $P(Z < z) = 0.20 \Rightarrow z = -0.84$.
(b) $P(Z \leq z) = 0.50 \Rightarrow z = 0$.
(c) $P(-z \leq Z \leq z) = 0.50 \Rightarrow z = 0.6745$
(d) $P(|Z| > z) = 0.01 \Rightarrow z = 2.576$
(e) $P(|Z| < z) = 0.90 \Rightarrow z = 1.645$

43. (a) $P(Z < -2.0537) = 0.02$, so the worst case percentage change is $0.08 - 2.0537(0.2) = -0.33074$. The investment could fall 33% in value, meaning that the value at risk is $33,000.
(b) For the value at risk to be reduced to $20,000, a 2% percentage change must be equivalent to a loss of 20%. Solving $\mu - 2.0537(0.2) = -0.2$ for μ implies that the growth would have to be approximately 21%.
(c) No, the value at risk does not add up over time this way because standard deviations don't add up either. Consider the conditions for "b." If $\mu = 0.21$ with $\sigma = 0.2$, this investment puts $20,000 at risk annually (at 2%). Ignoring compounding, for two years, the expected growth would be 42%. If the results are uncorrelated between years, the SD of the sum of the returns is $\sqrt{0.2^2 + 0.2^2} = \sqrt{0.08} \approx 0.28$. The two-year value at risk is then $100,000 \times (0.42 - 2.0537 \times 0.28) \approx 100,000 \times -0.155 = \$15,500$. The value at risk is *smaller* over the longer horizon (primarily because the mean return is so large relative to the SD).

45. (a) It loses on one policy with probability 0.025, the chance for a driver to have an accident.
(b) The insurer takes in $2.5 million for the 1,000 policies. At a fixed cost of $65,000 per accident, the insurer will lose money if more than $2,500,000/65,000 = 38.46$ accidents occur. That is, if 39 accidents occur. The probability of 39 or more accidents is precisely a binomial calculation that we can approximate using a corresponding normal distribution. Let $X \sim \text{Bin}(n = 1000, p = 0.025)$. The normal approximation is

$$P(X \geq 39) \approx 1 - P(X \leq 38) \approx 1 - P(Z \leq (38-np)/\sqrt{np(1-p)})$$
$$= 1 - P(Z \leq (38-25)/\sqrt{1000(0.025)(1-0.025)})$$
$$\approx 1 - P(Z \leq 2.633122) \approx 0.00423$$

If you have software that will do it, a binomial model gives the exact answer (under these somewhat artificial circumstances). The probability is $P(X \geq 39) = 1 - P(X \leq 38) \approx 0.00514$. The difference is small numerically, but large on a relative scale. If you have the software to do it, use the binomial distribution for these calculations that have small probability. (The normal calculations also vary depend on how you treat the inequalities.)
(c) Yes, to the extent that the company can be profitable by writing many policies whereas it would not if it only sold a few.

47. (a) 5%, and in this case it loses a lot!
(b) 5%. Either all of the bonds pay or they do not. These are not independent contracts. They all pay at the same time.
(c) The life insurance firm has independent customers, they don't all die at once. The hurricane bonds are not independent. The hurricane bonds are much more risky than life insurance.

Chapter 12 The Normal Probability Model

49. (a) The histogram and boxplot (shown below) look like a reasonable match for a normal model. The histogram is roughly bell-shaped and the boxplot does not flag many outliers. The histogram does, however, show some skewness and a sharp (rather than gradual) cutoff near 30%.
(b) Some slight outliers. The lowest share is 24.9% (in Jersey City, NJ); the highest is 43.7% (in Buffalo, NY).
(c) The shares of this product are never very close to zero or 100%, so the data do not run into the boundary at the upper and lower limits. The boundaries do not affect the distribution.
(d) Set the mean $\mu = 33.63\%$ ($\bar{x}$) and $\sigma = 3.34$ (s).
(e) The normal quantile plot suggests an acceptable match, but for the kink in the quantile plot near 30%. This cluster of values is unusual for data that are normally distributed. Otherwise, the normal model describes the distribution of shares nicely, matching well in the tails of the distribution.

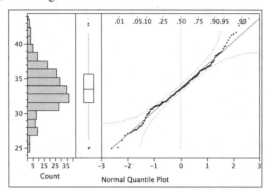

Chapter 13 Samples and Surveys

Mix and Match

1. c

3. a

5. i

7. j

9. f

True/False

11. True. A SRS makes all subsets of the indicated size equally likely.

13. False. Bias occurs when the sample is not representative. Sampling variation refers to the differences due to randomization among the samples.

15. False. The sampling frame is a list of the items in the target population.

17. False. Voluntary response surveys allow self-selection rather than random selection.

19. False. Examples described in the text, such as the polling numbers, have shown the opposite effect.

Think about It

21. Population: HR directors at Fortune 500 companies
    Parameter: Proportion who find that surveys intrude on workday
    Sampling frame: List of HR directors known to business magazine
    Sample size: Apparently 23% of 500, or 115
    Sample design: Voluntary response
    Other issues: The question is directly related to whether an individual participates. Presumably, those who do find these surveys intrusive do not have time to reply.

23. Population: The production of the snack foods manufacturer
    Parameter: Mean weight of bags (or perhaps the percentage underweight)
    Sampling frame: All cartons produced daily (presumably they have a list)
    Sample size: 10 cartons × 2 bags each
    Sample design: The cartons are a simple random sample of the production; the two bags sampled (randomly?) within the cartons form a stratified sample (each case is a stratum).
    Other issues: How are the bags chosen from a carton?

25. Population: Adult shoppers
    Parameter: Satisfaction (rating, percentage who request coupon)
    Sampling frame: None used; the subjects are those who visit the store.
    Sample size: Not given, but probably large if it's a hot day and the beverage is cold!
    Sample design: Convenience sample
    Other issues: This effort is more promotion than survey. The intent appears more to advertise the product than to learn customer opinions (though that might happen too).

27. (a) Voluntary response survey. Are on-line users representative of those who visit branches, or do they do all of their banking on-line and would not use either alternative?
    (b) Cluster sample/census. This will give a good idea of the reaction at this one branch, but customers at other branches might have different habits/preferences.
    (c) Voluntary response, with all of the problems of these surveys (e.g., not representative, only get responses from those with strong opinions)
    (d) Cluster sample. SRS within each branch with plans for follow-up.

29. The wording of Question 1 makes it sound as though the bank doesn't care about its employees and is likely to produce a different response than Question 2.

31. (a) If the payments are numbered and recorded electronically, take an SRS (using a spreadsheet program) of ten of them. If they are kept in a big stack of paper, pick a random integer from 1 to 20. Then pick every 10th payment to check. The SRS is better, but it might be easier to get someone to follow the paper procedure.
(b) Separate the two types of payments and perhaps have different supervisors sample each. If done by the same person, have that person pick five of each (or proportionally if many more of the payments are of one type).

Chapter 14 Sampling Variation and Quality

Mix and Match

1. The x-bar charts in (a) and (b) signal that the process goes out of control (top left, in group 16, and top right, going too low also on about day 16).

3. (a) S goes out of control (too high) in group 13.
   (b) In control, though S gets close to the upper limit in group 13.
   (c) The X-bar chart goes out of control in group 7 (albeit just barely). S is trending up in the later groups, but remains inside the control limits.
   (d) Out of control in both: first in the S chart in group 9 and then in the x-bar chart in group 10 as well.

True/False

5. True.

7. False. Stability in the process does not imply a Type I error. The error occurs if the system indicates a problem when in fact none has occurred.

9. True. $\sigma/\sqrt{n} = 5/\sqrt{25}$ $\sigma/\sqrt{n} = 1$ lb.

11. False. This default choice focuses our attention on a Type I error and effectively ignores the risk for a Type II error. The wide limits make it easy to miss a change in the process.

13. True. The process mean μ is also the mean of the sample averages.

15. False. The distribution of withdrawal amounts is evidently not normally distributed. Notice that the SD is nearly as large as the mean, but negative values are not possible.

17. False. The procedure indicates that the process is out of control. This may have occurred because the process changed or simply by chance. Only in the latter case does a Type I error occur.

19. True. Consider the control limits with $\alpha = 0.0027$ which are at ± 3 SE($\overline{X}$). The larger the sample size, the tighter these become with the same value of α.

Think About It

21. (a) The change decreases the chance of a Type I error and increases the chance for a Type II error because this procedure makes it more difficult to signal a halt to the procedure.
    (b) P(Type I) = 0.05^2

23. (a) Under control unless there's a special sale or weekend shopping surge.
    (b) Under control unless some problem causes a surge in calls.
    (c) Out of control with surges during holiday season.
    (d) Out of control with upward (or perhaps downward!) trend.

25. (a) Poisson, with rate $\lambda = 4$ units per hour. Verify that the counts in disjoint time intervals are independent. It would also be useful to see a histogram of production per hour when the process is running normally to compare to a Poisson model.
    (b) Let $Y \sim$ Pois($\lambda = 4$). The process signals a problem if the count in an hour is 0 or 1. The probability implied by the Poisson model when λ is the correct value is $P(Y = 0) + P(Y = 1) = \exp(-\lambda)(1+\lambda) = 5\exp(-4) \approx 0.092$.
    (c) Set $\lambda = 3$. Then the probability of correctly detecting a problem is $P(Y = 0) + P(Y = 1) = \exp(-\lambda)(1+\lambda) = 4e^{-3} \approx 0.20$. Hence, the probability of failing to detect the problem is 0.80.

27. (a) If the manufacturer only checks five at a time, the small sample size will make it hard to find the defect on average. By using a larger sample, it narrows the control range to make it possible to find a subtle problem.
    (b) If the problem is very large, it would be best to check the parts daily to catch the problem more quickly than waiting until the weekend.

29. (a) $10 \pm 3 \times 5/\sqrt{18}$
    (b) $-4 \pm 3 \times 2/\sqrt{12}$

Chapter 14 Sampling Variation and Quality

31. (a) $0.986573 = 0.9973^5$
    (b) $0.763101 = 0.9973^{100}$

You Do It

33. (a) By the Empirical Rule, $P(Z > 2) = 0.025$, or 0.02275 using 1.96 from the normal tables.
    (b) Using the Empirical Rule, $P(-2 < Z < 2) = 0.95$. The probability of 80 such values in a row is then $0.95^{80} = 0.0165$. If you test 80 in a row, you're almost certain to get one out of the limits at ±2 SE. For $P(-3 < Z < 3)$, the Empirical Rules gives 0.9973, so the chances for 80 in a row within these limits is 0.9973 80 = 0.8055.
    (c) Yes. The limits for each day are inside the control bounds.

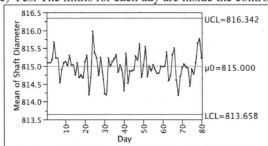
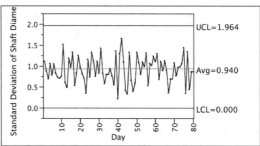

(d) No. With 2SE limits, the process appears out of control in the X-bar chart and the S chart.

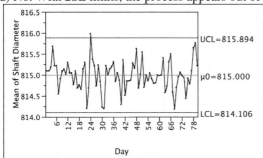
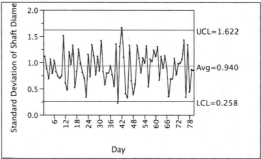

(e) If the process is under control, you are more likely to signal that the process is out of control when looking at the 80 daily summaries (part c) or the 16 weekly summaries (part d).

(f) The skewness and kurtosis using early data (such as the first 10 days before the charts signal a problem) are $K_3 = -0.14$ and $K_4 = -0.39$. Using 5 days is enough since n is larger than 10 times the square of K_3 or $|K_4|$. These data are very close to being normally distributed when we consider these summary measures.

35. (a) Using the first 5 days (60 observations, before the process got far out of control), the mean is about 449.88 with $s = 1.14$. The skewness during this period is $K_3 = -0.45$ (the data is left-skewed) and $K_4 = 0.81$. Ten per day is enough to meet the CLT condition, but just barely since $10 K_4 \approx 8$.
    (b) Use with $\mu = 250$ and $\sigma = 1.5$ (so that $\mu \pm 2\sigma$ holds 95% of the output). The process goes out of control. Values in both charts go outside the control limits.

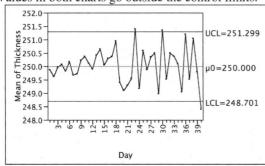
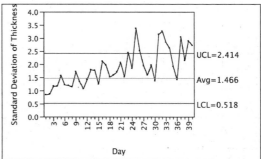

(c) The failure appears to be an increase in variation beyond that specified in the design. The standard deviation started out near 1, then hovered near the design target before drifting off. Because of the larger than designed variation, the mean values also appear to go out of control.

Chapter 15 Confidence Intervals

Mix and Match

1. f

3. b

5. d

7. j, assuming a sample survey for a proportion

9. e

True/False

11. True. Higher coverage means a longer interval.

13. True. The margin of error is proportional to $\sqrt{n}$.

15. True. A confidence interval identifies the collection of values for the population parameter that are consistent with the observed sample.

17. True. The 95% interval is a subset of the 99% interval.

19. False. The survey needs $1/0.05^2 = 400$. See Table 16.4.

Think About It

21. (a) [4.983 kg to 20.385 kg]
    (b) [¥267,490 to ¥511,720]
    (c) [$54.5 to $76.44]
    (d) [$18,600 to $29,160] per store

23. ½. At least approximately if the conditions of the z-interval for the mean are met.. The sampling distribution of $\overline{X}$ is centered at μ with half of the probability on either side.

25. You cannot tell which of these will be shorter unless you happen to know σ as well as s. If $\sigma = s$, the t-interval is longer because $z_\alpha < t_{\alpha,n-1}$ for any value of α. If the sample standard deviation is smaller than σ, however, the t-interval will be shorter for that sample.

27. (a) We can be 95% confident that the population mean height of men who visit this store (assuming the necessary conditions hold) lies between about 70.9 and 74.5 inches.
    (b) Half of the length, or $(74.4970 - 70.8876)/2 \approx 1.8$ inches.
    (c) Probably to the nearest inch.
    (d) Longer since the t-percentile would be larger.

29. The population average amount of sales is within $15 of the estimate, with some (probably 95%) degree of confidence.

31. (a) Use the expressions for the needed sample size from the text. To limit the error to 4% with 95% confidence we need $(1/0.04^2) = 625$.
    (b) This result requires manipulations similar to that producing the expression in the text. A 98% confidence interval is of the form ± 2.33 SE. Hence, we must have $2.33/2 \times 1/\sqrt{n} = 0.04$ or $n = 1.165^2/0.04^2 \approx 849$.

33. (a) PD × EAD × LGD = $0.06 \times 250000 \times 0.20 = \$3,000$
    (b) The expected loss becomes uncertain. Taking the approach used in the text to combine intervals, the uncertainty is $0.05 \times 220000 \times 0.18 = \$1,980$ to $0.07 \times 290,000 \times 0.23 = \$4,669$
    (c) If the estimates are independent, then the product has coverage of at least $0.95^3 \approx 0.86$. If the estimates are dependent, the interval may not have this much coverage, or it might have much more.

You Do It

35. (a) 1.796. Remember, $n-1$ df.
    (b) 2.571
    (c) 2.977

37. (a) Population: cars serviced at this dealership
    Sample: cars serviced during the time of the observations
    Parameter: p – proportion of all cars with these dents
    Statistic: $\hat{p}$ – proportion seen with these dents (25.3%)
    Issues: The sample is relatively small (only 87) and the context suggests that it may not be representative. Perhaps the cars brought in during this period are different from those at other times.
    Interval: $25.3\% \pm 1.96 \times \sqrt{\dfrac{0.253(1-0.253)}{87}} = [0.1616482, 0.3443518]$
    (b) Population: shoppers at supermarket;
    Sample: those who returned the form
    Parameter p: population proportion of those who find shopping pleasing
    Statistic: $\hat{p}$ - proportion of those who returned the form that find shopping pleasing (250/325)
    Issues: The methods of this chapter should not be used because of the voluntary nature of the sample. The sample is likely to be biased and non-random.
    (c) Population: visitors to web site;
    Sample: those who fill in the questionnaire
    Parameter: μ – average web surfing hours of all visitors
    Statistic: $\bar{y}$ - average hours of those who complete survey
    Issues: If this is indeed the relevant population (visitors to this one site) and these have indeed been sampled, they you can use the methods of this chapter. In general, Internet surveys have relative low quality due to non-response and the voluntary nature of the survey.
    Interval: $3 \pm 1.971 \times 1.5/\sqrt{223} = [2.802018, 3.197982]$ hours
    (d) Population: customers given loans during the past 2 years
    Sample: the 100 customers sampled
    Parameter: p – proportion of customers who default on loan
    Statistic: $\hat{p}$ - proportion of these 1000 who default (0.2%)
    Issue: You cannot use the z-interval of this chapter because the sample proportion is too small.

39. (a) Those in the checklist for the t-interval: That these data are a SRS from the "population" of revenue streams (a hypothetical population, as in control charts) and that the sample size is sufficient for the CLT condition to be met.
    (b) The interval is $1264 \pm 1.681 \times 150/\sqrt{44} = 1225.987$ to 1302.013. We'd round this to the dollar scale, as $1,226 to $1,302 (or perhaps 10s of dollars).
    (c) Just barely. The value $130 lies inside this interval.
    (d) Multiply the endpoints of the prior interval by 5, obtaining a rounded interval that goes from $1226 \times 5 = \$6230$ to $1302 \times 5 = \$6510.3$.

41. (a) $1.976 \times 8/\sqrt{150} = 1.290718$ (or use 2 standard errors rather than a t-quantile shown here)
    (b) They are 95% confident the average waiting time for all callers is within about 1.3 minutes of the average wait of 16 minutes found in this sample.
    (c) Smaller. Less confidence for a given sample allows a narrower interval.
    (d) $1.655 \times 8/\sqrt{150} = 1.081041$

43. (a) Not correct. The interval describes with 95% confidence the location of μ, not the individual shipments
    (b) Not correct. The interval does not describe the individual shipments
    (c) Correct.
    (d) Not correct. The interval does not describe the mean of another sample.
    (e) Not correct. This is the wrong interval.

Chapter 15 Confidence Intervals

45. (a) se = $35/\sqrt{60} \approx 4.518$, $t_{0.025,59} \approx 2.001$
    t-interval $152 \pm 2.00 \times 4.5185 = [142.963, 161.037]$.
    (b) se = $75/5 = 15$, $t_{0.025,24} \approx 2.0639$
    t-interval $8 \pm 2.064 \times 15 = [-22.9585, 38.9585]$.
    (c) se = $0.5 \times \sqrt{1/75} \approx 0.0577$
    z-interval $0.5 \pm 1.96 \times 0.0577 = [0.386908, 0.613092]$.
    (d) se = $\sqrt{0.3(0.7)/23} \approx 0.0956$
    z-interval $0.30 \pm 1.96 \times 0.0956 = [0.112624, 0.487376]$.

47. The confidence interval is $123/1000 \pm 1.96 \sqrt{\hat{p}(1-\hat{p})/1000} = (0.1026433, 0.1433567)$. We are 95% confident, based on this sample, the proportion of people contacted who may buy something is between 10.3% and 14.3%.

49. (a) Though you cannot be sure (miracles do happen), the interval for the manufacturer is most likely much shorter because it has a much larger sample and a correspondingly smaller value for the standard error.
    (b) This is not the correct interpretation of a confidence interval.
    (c) No, both are 95% intervals.

51. This would require a margin for error of less than 0.01. Using the formula ME = $1/\sqrt{n}$, the survey would need to have at least $1/0.01^2 = 10,000$.

53. (a) If this week is typical of the activity, then we could think of these 1,200 as representing a sample of all clicks to this company. Certainly, it would not be appropriate to do this if the chosen week were not representative, such as during a holiday season or during a special promotion.
    (b) The z-interval is appropriate here because of the large n (both np and n(1-p) are big.) $\hat{p} = 175/1200 \approx 0.1458$ with se $\approx \sqrt{0.1458(1-0.1458)/1200} \approx .0102$ Hence the 95% confidence interval is
    $0.1458 \pm 1.96 \times .0102 = [0.1258$ to $0.1658]$
    or about [0.126 to 0.166].
    (c) At $4.50 per click, with 0.1458 fraudulent, we simply multiply the prior interval by 4.5. We'll round to nearest penny in this interval.
    $[0.1258$ to $0.1658] \times 4.5 = [.5661$ to $.7461]$
    or about [$0.57 to $0.75] in costs due to fraud per click.

Chapter 16 Statistical Tests

Mix and Match

1. i

3. h

5. a

7. e

9. b

True/False

Questions 11-16: $H_0: \mu \leq 80$, $H_a: \mu > 80$

11. False.

13. True.

15. False. The smaller the α-level the less likely a test is to reject H_0.

17. False. When σ is estimated by s, the test statistic is a t-statistic.

Questions 17-24: $H_0: p \leq 0.4$, $H_a: p > 0.4$

19. False. Maybe, but not necessarily. Sampling variance could produce a sample in which $\hat{p}$ is more than 0.4 even though $p \leq 0.4$.

21. False. A statistically significant result will only occur if $\hat{p}$ is large enough in order to show that more than 40% will use the service. The z-statistic will need to be larger still. (*e.g.*, If $\hat{p} = 0.400001$, the test would not reject H_0 unless n were huge.)

23. True.

25. False. The *p*-value is the probability of incorrectly rejecting H_0 and adding a service that will not be profitable.

Think About It

27. The null hypothesis is that the therapy is not effective. That is, the test must demonstrate effectiveness of the medication. Hence, $H_0: \mu \leq 10$ mm vs $H_a \mu > 10$. (In many cases, however, the Food and Drug Administration requires the use of two-sided tests nonetheless.)

29. Based on the context, the managers should be worried about missing a blockbuster product. That would typically mean a Type II error. That said, if the company tests millions of products (a common occurrence in automated material science), managers might also be wise to worry about Type I errors as well. (See the next question.)

31. (a) If $p = ½$, then we get all heads with probability 1/8 and all tails with probability 1/8. Hence, the chance for incorrectly rejecting H_0 is ¼.
    (b) $P(3 \text{ heads}) = (3/4)^3 = 27/64$. $P(3 \text{ tails}) = (1/4)^3 = 1/64$
    Hence, $P(\text{reject}) = 28/64 = 7/16$. Since H_0 is clearly false, the Type II error is the probability that we do not reject, or 9/16.

33. (a) $P(\text{convict innocent defendant}) = 1/2^{12} \approx 0.00024$
    (b) Type I error, false positive
    (c) $P(\text{fail to convict guilty}) = 1 - 0.95^{12} \approx 0.46$
    (d) Type II error, false negative

35. Evidently the study sample size was unnecessarily large (and hence the study was far too expensive). The drug was quite good and could have been shown as successful with a much smaller sample.

37. (a) Set $H_0: \mu \leq 200$ so that the score "proves" that an applicant's true mean is above 200.
    (b) Using the Empirical Rule as a rough approximation to the normal distribution the HR group rejects H_0 if the

Chapter 16 Statistical Tests 49

score is larger than $200 + 2 \times 25 = 250$. (Essentially $n = 1$ here, so σ is the standard error.)
(c) If μ is 225, this mean lies one SD below the threshold 250 defined in "b". Hence, (again using the Empirical Rule) there's about a 1/6 chance to reject H_0 and the chance for a Type II error is about 5/6. Even though H_0 is false, there is little chance to identify the problem from a single measurement.

You Do It

39. (a) Yes, reject H_0 at the usual α level of 0.05 (or even smaller). The number of damaged washers found is more than expected. Let the r.v. X denote the number of damaged washers among the 60 that are inspected. If $p = 0.02$ (the upper limit allowed for p under H_0), then the probability of finding 5 (or more, allowing more extreme deviations from H_0) is
$$P(X \geq 5) = 1 - P(X < 5) = 1 - (P(X = 0) + P(X = 1) + \ldots + P(X = 4))$$
$$\approx 1 - (0.298 + 0.364 + 0.219 + 0.087 + 0.025) = 0.007$$
That's the p-value of the test, and is small enough to reject H_0 at the usual $\alpha = 0.05$ level.
(b) The inspection of the washers must produce a sequence of independent outcomes with constant chance p for spotting a defective item.
(c) If we directly apply the z-test, then the observed proportion $5/60 \approx 0.0833$ is
$$z = (0.0833 - 0.02)/\sqrt{0.02(0.98)/60} \approx 3.5 > 1.645$$
standard errors above H_0. This test also rejects H_0 at the usual $\alpha = 0.05$ level.
(d) Although both tests reach the same conclusion (reject H_0), the sample size condition is not satisfied in this application. The expected number of events under the null hypothesis (2% of 60 = 1.2) is too small for the Central Limit Theorem to be effective. Hence, a normal model is not a good description of the sampling distribution of the sample proportion. You have to go back to binomial methods for this one. (The p-value of the test is 0.0002, far smaller than the exact calculation given by the binomial in (a).)

41. (a) H_0 $p \leq 0.6$, where p is the proportion of markets selling out in this market. The alternative hypothesis is H_a: $p > 0.6$
(b) A Type I error occurs if we reject H_0 incorrectly, adding an unnecessary delivery. A Type II error occurs if we fail to reject H_0 when it's false, missing an opportunity.
(c) The sample proportion is $35/45 = 7/9$. The standard error is $\sqrt{0.6(0.4)/45} \approx 0.07303$. Under H_0: $z = (7/9 - 0.6)/0.07303 \approx 2.434$. Using the rounded z value and the table, the p-value is approximately $P(Z > 2.43) = 0.0075$.

43. (a) H_0: $p \geq 0.33$, where p is the proportion of all visitors who indicate in the survey a willingness to return to the hotel. The alternative hypothesis is H_a: $p < 0.33$
(b) A Type I error occurs if we reject H_0 incorrectly, intervening at the local franchise unnecessarily. A Type II error occurs if we fail to reject H_0 when it's false, missing the opportunity to intervene and correct a problem.
(c) The sample proportion is 0.2. The standard error is $\sqrt{0.33(0.67)/80} \approx 0.052$. Under H_0, $z = (0.2 - 0.33)/0.052 = -2.5$. Because $z < 1.96$ (or use 2 from the Empirical Rule), we reject H_0. The p-value is approximately $P(Z < -2.5) = 0.0062$.

45. (a) H_0: $\mu \leq \$120$, where μ is the average amount spent by a shopper who participates in the loyalty program. The alternative hypothesis is H_a: $\mu > 120$.
(b) A Type I error occurs if we reject H_0 incorrectly, that loyal shoppers spend more on average when in fact they do not. A Type II error means that loyal shoppers do spend more, but we did not reject H_0.
(c) $n > 10 K_4$, so K_4 must be less than 8.
(d) The test statistic is $t = \dfrac{130 - 120}{40/\sqrt{80}} = 2.236$. The p-value is about $P(T > 2.236) \approx 0.0140$. Reject H_0 since the p-value is less than $\alpha = 0.05$.

47. (a) Yes. These shoppers, being related, most likely affect the amount purchased by each other. The data would be dependent.
(b) No. The point of the program is to increase the amount of spending. Setting aside the outliers removes the shoppers that make the program profitable. Unless these large amounts are confirmed errors, use them.

49. The test of shoppers' spending rejects H_0 if the sample average lies more than 1.645 SEs above H_0. This occurs if the sample average is more than $120 + 1.645 \times 40/\sqrt{80} \approx \$127.36$. (The test in that question does reject.)

The chance of rejecting is given by the probability that an average of a sample from a population with $\mu = \$135$ being larger than $127.36. This is P($\overline{X} > 127.36$) = P(Z > (127.36 − 135)/$\left(40/\sqrt{80}\right)$ = -1.708) ≈ 0.956.

Hence, there's a 96% chance of rejecting H_0 correctly if $\mu = 135$, making the chance of a Type II error about 4%.

51. (a) H_0: $\mu \leq \$50$, where μ denotes the average increase in interest profit on a savings account when offered this personalized service (4% of the average balance in the account). H_a: $\mu > 50$ (and hence profitable on a larger scale).
(b) A Type I error means that the bank rolled out the program, but it will not be profitable. A Type II error means that the bank should have rolled it out (rejected H_0) but did not.
(c) The skewness must be less than $\sqrt{65/10}$ and the kurtosis must be less than 6.5 in order for the data to meet the CLT condition.
(d) The average increase in the balances is $1500, earning $0.4 \times \$5000 = \$60$ profit. The SD of this gain is $0.04 \times 3000 = \$120$. The test statistic is $t = (60 − 50)/(120/\sqrt{65}) = 0.6719$. The p-value is larger than 0.05. P(Z > 0.67) ≈ 0.25. Do not reject H_0 since the p-value is larger than α.

Chapter 17 Alternative Approaches to Inference

Mix and Match

1. g
3. b
5. f
7. j
9. c

True/False

11. False. The Central Limit Theorem that with enough observations (CLT condition), we can use a *t*-interval..
13. False. These checks are not needed.
15. False. This is a prediction interval for a new draw from the population.
17. False. Add a total of 4 cases, half success and half failure.
19. True. In more advanced courses, however, this claim would in general be false because there are population models for which the median exists, but not the mean. For us, those don't happen.

Think About It

21. Even though the data are symmetric on the log scale, these data should be checked for normality before using the *t*-interval. Plus, the interval on the log scale covers the mean of the log of the data. That's not the same as $\log \mu$, so converting back to dollars does not fix the problem (log(average) ≠ average(log) as in the text example.)
23. The data on the left appears to "touch" the boundary, but it is not clear. The plot on the right is okay. Both appear nearly normal, though it's a close call on the left. With small samples, it is very hard to recognize samples from a normal population.
25. No. Both intervals are based on approximating the actual sampling distribution of a proportion. The adjusted interval is shorter, but it also has a different location. If either has 95% coverage, it's the adjusted interval.
27. ½. That's what it means to be the median.
29. The adjusted small sample 95% interval for the proportion based on a small sample is more likely to include zero because of the shift in the value of the estimated proportion unless $\hat{p}$ is very close to 0. The reason that it's not always closer is that the adjusted interval is also a little shorter which makes it less likely to catch 0.
31. A prediction interval, since the retailer wants an interval for a single store, not an interval that describes a parameter of the population.
33. The longer interval is the prediction interval.
35. No. The intervals use the observed data which are all positive.

You Do It

37. (a) Yes, in spite of the outlier. According to a normal quantile plot, these data could be a sample from a normal population.

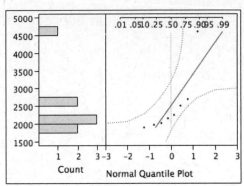

(b) No. The kurtosis, for example $K_4 = 5.8$, implying that we'd need a sample of $n = 58$ ore more to rely on the CLT in this situation.
(c) With: $1763 to $3245   Without: $1934 to $2478
(d) The mean falls from $2504 down to $2206, and the SD falls from $886 down to $294. The change in the SD is much larger than the change in the mean. The interval's center shifts down but the interval is also much shorter. The lower endpoint of the interval is higher than before.

39. The coverage probability is $1 - 2/2^8 = 1 - 1/2^7 = 127/128 \approx 0.9922$

41. Use the small-sample interval for the proportion. The adjusted proportion is $\tilde{p} = 12/18 = 2/3$ with standard error $\sqrt{\dfrac{2/3(1/3)}{18}} \approx 0.11$ The associated 95% interval is then $0.667 \pm 2(0.11) = 0.667 \pm 0.22$ which includes ½. This small sample allows that p in the population might be ½.

Chapter 18 Comparison

Mix and Match

1. j

3. c

5. g

7. a

9. e

True/False

11. True. If we reject H_0, then we want to be able to conclude that we've found a profitable procedure.

13. False. A test has a chance for making a false positive error. In the standard test, this chance is limited to 5%.

15. False. The difference $\mu_1 - \mu_2$ *might* be zero, but it does not have to be zero.

17. False. The boxplots may overlap but the difference between the means may be significant. See the weight-loss example.

19. True.

Think About It

21. Her choice might make sense, but not because of data. With only one observation of the time to commute by transit, she may have happened to hit a slow day. We also don't have a sense of the variation in the time to drive; you can be sure that it does not take exactly 40 minutes each day. Seems like a hasty decision to us.

23. Because of the pairing, this problem reduces to a one-sample analysis. Form the differences (such as "before" and "after") and compute the confidence interval for the difference.

25. (a) Simply take 40% of the two endpoints, [$200, $880] profits.
    (b) Yes, we can conclude that the methods used by Group A sell more. Because of the randomization, we have ruled out confounding factors as an explanation.
    (c) The randomization makes this bias unlikely, though we might want to compare the history of the sales people in the two groups to see whether in fact those in Group A sold more than those in Group B in the past.

27. These data mimic the advantages of an experiment. We can compare the change in energy consumption from 2005 to 2006 in homes that were on daylight savings time to the change in homes that moved from standard time to daylight savings time. It's important to compare changes since 2005 or 2006 might be an unusual year, weather-wise. By comparing the change in these two groups, we are using the homes that use daylight savings time in both years as a control group, setting a baseline for comparison. (For the results, see the presentation by M. J. Kotchen and L. E. Grant (2008) "Does Daylight Saving Time Save Energy? Evidence from a Natural Experiment in Indiana" which was discussed in The Wall Street Journal (February 27, 2008). They found that the switch to daylight savings time increased energy use by a statistically significant amount.)

You Do It

29. **Wine**

The data are numerical and labeled as a sample, but it is unclear about the population. It is unlikely that the wine seller would offer poor wines from any year. As a result, we're only going to see wines that get top ratings in any one year. The other conditions seem okay, but this flaw in the underlying sampling seems fatal. We'll do the rest of the calculations for practice, just the same.
(b) Our software provided the analysis shown below. The interval includes zero, suggesting no consistent preference from one year to the next. 2001 looks better, but could be explained as simply sampling variation.

53

2001-2000, Allowing unequal variances

Difference	0.6880	t Ratio	0.93626
Std Err Dif	0.7348	DF	31.67736
Lower CL Dif	-0.8094	Prob > |t|	0.3562
Upper CL Dif	2.1854	Prob > t	0.8219
Confidence	0.95	Prob < t	0.1781

(c) The comparison would be nonsensical, if for example, all of the wines from 2000 were white wines and all from 2001 were red wines. Other potential sources of confounding include the sources of the wines, such as differences in nationality.

(d) The se of the difference rounds to 0.7, so the interval should be presented as 2.2 down to -0.8.

(e) The confidence interval might include zero, but wines from the 2001 vintage score higher on average. Unless there's a reason to stay with 2000, common sense says the 2001 bottle is better.

31. **Used Cars**

There are substantial differences in variation between the two groups, but since we are using a method that does not require the variances to be identical, we can proceed. The 95% confidence interval for the difference in the means (I – Xi) is shown in the output below. The interval does not include zero; indeed, the Xi model sells for about $600 to $2,800 more, on average.

(b) No. The average age of the cars in the two groups is identical. Age has not confounded the comparison in (a). (Had, for example, the Xi models been newer, then this difference in age might have explained the difference in prices.)

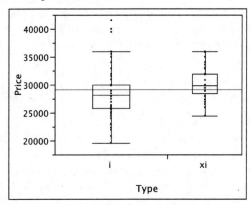

Difference	-1698.9	t Ratio	-3.10693
Std Err Dif	546.8	DF	152.9635
Upper CL Dif	-618.6	Prob > |t|	0.0023
Lower CL Dif	-2779.3	Prob > t	0.9989
Confidence	0.95	Prob < t	0.0011

Chapter 19 Linear Patterns

Mix and Match

1. e

3. f

5. b

7. a

9. c

True/False

11. False. The response y is drawn on the vertical axis with the predictor x on the horizontal axis.

13. True. The correlation is either +1 or -1. In either case, $r^2 = \text{corr}^2 = 1$.

15. True. Linear means equal differences in y come with equal differences in x. Of course, we should worry about extrapolating too far from the observed data.

17. False. The slope indicates change, not the intercept.

19. False. The residual is the *vertical* distance between these values, measured in the units of the response.

21. False. The residual plot should be simple, with the residuals clustered near the horizontal line at zero.

Think About It

23. No. Square the correlation to estimate the common variation between x and y. These two share ¼ of their variation in common.

25. No. Least squares criterion minimizes vertical deviations. In the reversed regression, the deviations change to those that were horizontal in the plot of y on x.

27. $s_e = \$32$. If the data are bell-shaped, then $\pm 2\, s_e$ holds about 95% of the residuals. 25 looks too small and 50 seems too large.

29. The slope becomes steeper, producing a higher cost per mile driven. The intercept would increase as well.

31. They are the same.

33. (a) The average purchase goes up by about $85 per shopper. A linear equation should capture this dependence in the plot. The equation would have a constant plus 85 times the number of shoppers.
(b) The intercept would be about 0, but is likely to be a large extrapolation from the rest of the data (unless this store does very little business). No shoppers mean no sales.
(c) The slope would be about 85 ($/shopper)
(d) Probably not. We expect that the variation would increase. Days with many shoppers might have either cash-heavy shoppers or, late in the month, shoppers watching their budgets.

35. (a) The intercept is $47,000 as a baseline for a store with no square feet. The intercept is most likely an extrapolation from the data and thus not directly interpretable. The slope is $650 per square foot. Because both x and y are in thousands, this common scaling factor (1000) cancels. For every one square foot increase, average annual sales increase by $650.
(b) The intercept $47 thousand would become $47 \times 0.82 = €\, 38.54$ thousand. The slope $650/square foot becomes
$$650\frac{\$}{\text{sq ft}} \times \frac{0.82 \text{ euros}}{\$1} \times \frac{1 \text{ sq ft}}{0.093 \text{ sq meter}} = 5{,}731\frac{\text{euros}}{\text{sq meter}}$$
We find that arranging the dimensions in this way keeps the conversions in the right position.
(c) r^2 is the square of the usual correlation, and does not depend on the units of the analysis. r^2 remains the same.
(d) s_e, which is measured in thousands of dollars, would change with the conversion to euros. To get the new value for s_e in euros, multiply the value in dollars by 0.82.

You Do It

37. Diamond Rings

(a) Yes, the scatterplot (shown below) shows a very linear trend.

(b) The least squares equation for the shown line in the figure is

Estimated Price (Singapore dollars) = -259.6259 + 3721.0249 *Weight*

The intercept is quite an extrapolation. One could have expected the intercept to be positive, accounting for the cost of the setting of the ring. Instead, it's -260 $S. The slope is 3721 $S/carat, which is also a bit of an extrapolation seeing the sizes of these stones are well under ½ carat.

(c) $r^2 = 0.9783$, indicating that the fitted equation describes all but about 2% of the variation in prices. The residual SD is thus quite small, s_e = $S 31.84. The data lie close to the line given the scale of the plot.

(d) This is 1/10 of the slope, or 372 $S/carat.

(e) 3721 $S/carat × 0.65$US/$S ≈ 2420 $US/carat. The slope for the emerald diamonds in the chapter is slightly higher at 2670 $US/ carat.

(f) One would expect the setting to add a fixed cost to the price. It may also be the case that the settings used for larger diamonds is more expensive than those for smaller diamonds. If so, there's a lurking variable behind this relationship.

(g) The residual for this ring is 325 - (-260 + 3721 × 0.18) = -84.78 $S

Because the cost is less than we expect for this size stone, the ring might be a bargain. Then again, it might not be such a nice gem (poor cut or low clarity).

(h) There's no evident pattern in the residual plot, suggesting it is okay to summarize these with a histogram (*i.e.*, simple variation). The standard deviation of the residuals is s_e = $S 31.84. Using the Empirical Rule, we'd expect about 95% of rings to be within $S 64 of the fitted line.

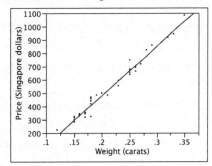 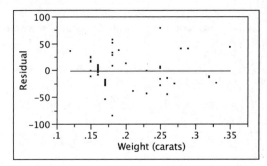

39. Download

(a) A line captures the general trend, though the fit of the data around the line is quite loose, with substantial residual variation.

(b) *Estimated Transfer Time* (sec) = 7.2746633 + 0.3133071 *File Size* (MB)

The intercept, 7.27 seconds, estimates the amount of "latency" in the network that delays the initial transfer of data regardless of the file size. The slope is the transfer rate, 0.3133 seconds per megabyte.

(c) $r^2 = 0.6246$ and $s_e = 6.2433$. The fitted equation describes about 62% of the variation in transfer times, with the remaining variation left in the residuals, which have SD 6.2433 seconds. These seem to be reasonable summaries in this example.

(d) By converting the units of both y and x, the colleague gets a rather different looking equation, but one that fits the data equally well (same r^2). b_0 becomes smaller 7.2746633 sec × (1 min)/(60 sec) = 0.1212 minutes; b_1 becomes 0.3133071 (sec/MB) × (1 min/60 sec) × (1 MB)/(1024 KB) = 0.0000050993 minutes per kilobyte.

(e) The residual variation lacks patterns. The vaguely evident "columns" of points in the plot are present because among these 80 files, the sizes were roughly bunched. For example, several files were of about 60 MB or 70 MB. Almost none were about 80 MB.

(f) To make the problem easier, do the regression in reverse. If we need to predict the file size that can be on average transmitted in a fixed length of time, change the roles. The resulting equation is

Estimated File Size (MB) = 6.8803225 + 1.993418 *Transfer Time* (sec)

Now we can easily see that given a 15 second window, we could expect to transmit a file of 6.8803225 + 1.993418 × 15 ≈ 36.78 MB.

Chapter 19 Linear Patterns

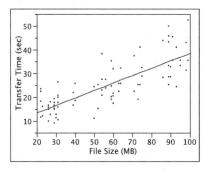
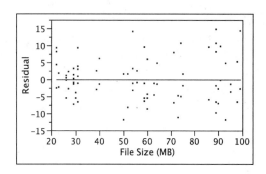

41. **Seattle homes**
    (a) The scatterplot shows a linear trend, with increasing variation.
    (b) *Estimated Price ($000) = 73.938964 + 0.1470966 Square Feet*
    The intercept is presumably the average cost of the land that the homes sit on (as if the lots are the same size). The slope is the cost (in thousands of dollars) per square foot, and so more naturally expressed as $147 per square foot.
    (c) $r^2 = 0.5605$, implying that the linear trend describes slightly more than 56% of the home-to-home variation in prices. The standard deviation of the residuals is $s_e = \$86.69$. Almost half of the variation in prices is due to other factors aside from size alone.
    (d) This equation suggests that adding 500 square feet might be worth an additional $500 \times 147 = \$73,500$ on average. We cannot be sure, however, that other factors do not lurk behind this relationship. Perhaps the larger homes are in more desirable areas, and we will not see this increase in value with the addition to the home.
    (e) The residual price for this home is (in thousands of dollars)
    $$625 - (73.938964 + 0.1470966 \times 2690) \approx \$155 \text{ thousand}$$
    This home costs more than most of this size. This residual is almost twice the size of the standard deviation of the residuals s_e. Perhaps this home is in a nice location with a great view.
    (f) The residuals show a fan-shaped pattern. The variation around the fit increases steadily with the size of the home. It would not be appropriate to summarize this variation with a single value.

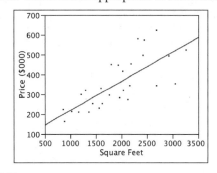
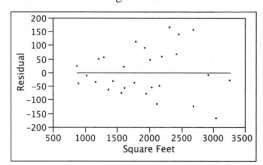

43. **R&D Expenses**
    (a) The scatterplot (shown below) shows a very strong linear trend if we accept two large outliers (Microsoft and Intel).
    (b) The least squares equation for the shown line in the figure is
    Estimated R&D Expense = 3.9819 + 0.0869 Assets
    b_0 estimates that a company with no assets would still spend $3.98 million on research and development. Although we have quite a bit of data near zero, this seems to be quite an extrapolation. Perhaps more reasonable is to interpret the intercept as a "commitment" to research that stays the same regardless of fluctuations in assets. The slope indicates that these companies on average spend about 9 cents out of each additional dollar in assets on R&D.
    (c) $r^2 = 0.9521$ with $s_e = \$98.2915$ million. The large size of r^2 above 95% appears to reflect the fit to the large outliers rather than the variation in the smaller companies. The scatterplot "squishes" small companies into the lower left corner of the plot.

(d) Both histograms are skewed. This skewness anticipates the outlier-dominated scatterplot. Several large companies dominate both the histograms and scatterplots. (Large values in one histogram are paired with large values in the other.)

(e) The residuals show a pattern, with most of the 504 cases bunched near zero on the left. The variation appears to increase rapidly with the assets of the companies. Because the variation changes, a single summary like s_e is inadequate.

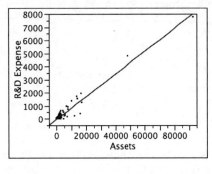 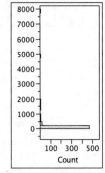 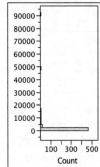

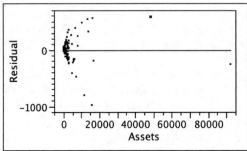

45. **OECD**

(a) The association is positive, of moderate strength, and vaguely linear. Conservative economists probably like the direction, with exporting economies generating larger GDP per capita.

(b) *Estimated GDP* (per cap) = 26,714.45 + 1,440.51 *Trade Bal* (%GDP)

The intercept is not an extrapolation! b_0 indicates that a country with balanced imports and exports has on average $26,714.45 GDP per capita. b_1 indicates on average countries with one percent higher trade balances (as a % of GDP) have $1,441 more GDP per capita.

(c) $r^2 = 0.369$, meaning that the line "explains" about half of the variation among countries in GDP per capita. s_e = $11,335.8 GDP per capita is the standard deviation of the residuals around the trend.

(d) The residuals appear random, but with so few it is hard to detect all but the most extreme patterns. There might be higher variation among the importing countries (negative trade balances).

(e) Norway has the largest GDP and Ireland the largest trade balance.

(f) The US (marked with a + in the plots, toward the left) is a net importer, but also has high GDP per capita given its level of imports (4.5% of the GDP). The residual for the US is

   39700 – (26,714.45 + 1,440.51 × (-4.5)) = $19,467.85

GDP in the US is more than $20,000 per person higher than anticipated by the fit of this equation.

Chapter 19 Linear Patterns 59

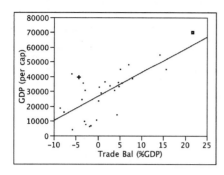
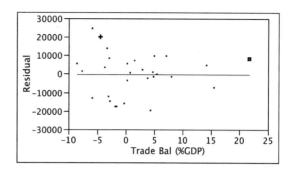

47. Promotion

(a) The timeplots (below) suggest little association between the variables (not that it's easy to tell from this view of the data). The market share seems to be declining over these weeks, but the trend if any is very slight. There's also not much pattern in the detailing voice.

(b) The scatterplot reveals some association, though weak. It is hard to identify this association from looking at two time series, side by side. The scatterplot is much more effective. A line seems to be a reasonable summary of the association that is present.

(c) *Estimated Market Share* = 0.211254 + 0.13005 *Detail Voice*
b_0 estimates that with no detailing, this drug would capture about 21% of the market. The slope indicates that on average, weeks in which this drug has 1% more of the promotion, it has 0.13% more share of the market. (Increases in promotional share do not come with equal gains to sales.)

(d) $r^2 = 0.14178$; detail voice describes only 14% of the variation in market share over these weeks. $s_e = 0.0071$; the SD of the residuals around the fitted line is 7.1% (large, considering that market share is about 20%).

(e) An increase in voice from 4% to 14% would on average increase share by $0.1 \times 0.13 = 1.3\%$. (This doesn't seem worth it unless share in the market is worth a lot more than increasing the voice costs.)

(f) The residuals appear random, albeit with one straggler (week 6) drifting off to the left of the plot.

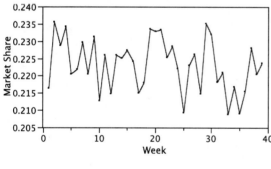
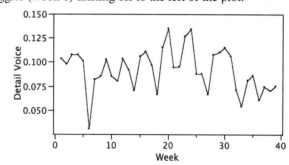
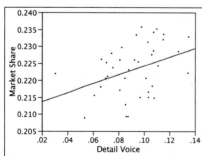
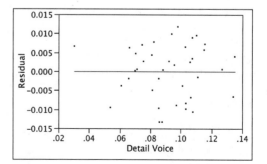

Chapter 20 Curved Patterns

Mix and Match

1. d
3. a
5. e
7. j
9. h

True/False

11. False. Transformations capture curves.

13. True

15. True. If the variable is measured as a ratio of "something per something else", a reciprocal flips this over. It's still a ratio.

17. False. Transformations are intended to capture patterns in the data, not make outliers look "more natural." Often, outliers do seem more reasonable after a transformation, but this is more often an accident.

19. True

Think About It

21. If the relationship is linear, these would cost the same amount if there are no (or small) fixed costs. For example, if diamonds cost $2,500 per carat with no fixed costs, the fitted line is $\hat{y} = 2500$ *Weight*. A one-carat diamond then goes for $2,500 and a ½ carat diamond for $1,250.

23. Heavier cars also have more powerful engines on average. Larger engines burn more fuel.

25. They are unrelated. Changes in the price would have no effect on the quantity sold.

27. b_0 is the fitted value when x gets large (the asymptote).

29. The stripes come from the practice of reporting gasoline mileage in whole numbers (rounding).

You Do It

31. **Wal-Mart**
(a) The scatterplot follows. The pattern is not linear, and grows more rapidly in the later years.
(b) The scatterplot shows the fit of the linear equation,
$$\text{Estimated Operating Income} = -570{,}031.1 + 286.35664 \text{ Date}$$
The intercept is a gross extrapolation (to the year zero). The slope indicates that on average operating income is growing about $286 billion annually at Wal-Mart over these 16 years.
(c) The residuals show the up-down-up pattern common when a nonlinear pattern has been estimated using a linear equation. The residuals also show more variation in the later years, with wider swings around the trend.
(d) The scatterplot on a log scale is more linear, with more consistent variation over these years. The plot does, however, seem to "bend back" the other way from the initial plot. The log scale emphasizes the gyration in sales in the late 1990's that is otherwise less apparent.
(e) The fitted equation is
$$\text{Estimated Log Op Income} = -306.7946 + 0.1572509 \text{ Date}$$
The slope (times 100) in this equation estimates the annual percentage change in operating income, here about 16%. This interpretation comes from algebra similar to that shown in the text. The estimated change in the log income from year t to year t+1 is
$$0.157 = \log y(t+1) - \log y(t) = \log(y(t+1)/y(t))$$
$$\approx \log(1 + (y(t+1)-y(t))/y(t))$$
$$\approx (y(t+1)-y(t))/y(t)$$

Chapter 20 Curved Patterns 61

(f) The residuals have an odd wavy pattern, tracking over time. We can also see that one quarter has exceptionally larger values over these years, namely the 4th quarter associated with the holiday shopping season.
(g) The log transformation reveals some details that are otherwise less noticeable (seasonal pattern, dip in sales in the late 90's). The slope also has a nice interpretation as the rate of growth. We cannot compare r^2 or s_e since the response in these two equations differs.

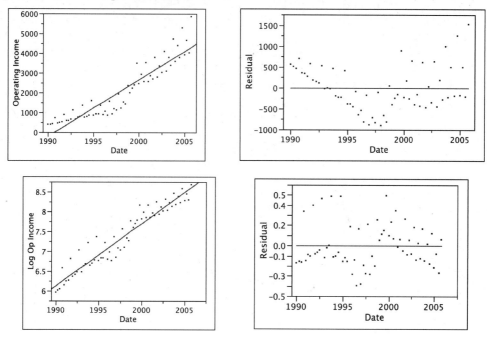

33. **Wine**
(a) The scatterplot follows. The relationship seems fairly linear over this range of ratings, though the floor of prices at zero complicates the analysis.
(b) The linear regression is shown in red. The equation of the fit is
  Estimated Price = -1058.588 + 12.184283 *Rating*
with $r^2 = 0.55$. The plot shows that the linear model under predicts poorly rated wines (they're not giving them away) as well as those that are very highly rated.
(c) The plot on the log scale appears to have a more linear trend than the original scatterplot.
(d) The fit of the model on the log scale is *Estimated* log(*Price*) = -19.67249 + 0.2566828 *Rating*
As shown in plot below, the log model (curve) does not predict negative prices for lower rated wines.
(e) We cannot use r^2 and s_e to compare these models because the response in the two models is different. The scale of s_e is not the same in the two models. Similarly, we should not compare directly explaining variation in the log of price with explaining variation in the price itself. We'll stick to the plot and the substantive common sense of preferring a model that does not predict negative prices for common situations.

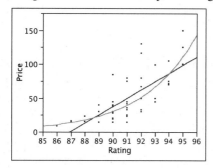 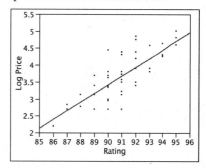

35. **Used Accords**
(a) It would be odd if the price fell off at a constant rate. The values cannot go negative, after all. One should expect larger drops in the first few years.
(b) The fit of the linear equation (shown in the figure below as the red line) is
$$\text{Estimated Asking price} = 15.462647 - 0.9464141 \, Age$$
The intercept suggests a new "used" car (just driven off the dealer's lot) is about $15,463. The slope indicates these cars drop in resale value by about $946 per year.
(c) The plot of the residuals (below left) shows that this equation misses the pattern, underestimating the price at the two ends and overestimating the price in the middle of the age range.
(d) These residuals appear more random when compared to the pattern that is evident in the residuals from the linear equation.
(e) The log predictor implies that price changes rapidly among relatively newer used cars, then drops off more slowly as the cars age.
(f) We can compare these summaries in this example because the response variable is the same. The log equation fits better, with larger r^2 and smaller s_e. By bending, the log equation captures more of the variation in the data. This table shows the results.

Model	r^2	s_e
Linear	0.795	$2,190
Log	0.928	$1,300

(g) The intercept is the predicted asking price for a car that is one year old, $22,993. Using the same approximations as in the text, we see that a 1% increase in age comes with a change in the resale price of about $0.01 \times b_1 = 0.078$ thousand dollars. This implies a much larger change in the first years of aging.
h) The estimated asking price drops almost $5,500 from $22,993 at age 1 to 17,573 in year 2. For older cars, the drop is smaller. The estimated price drops from $4,243 at year 11 to $3,562 in year 12.

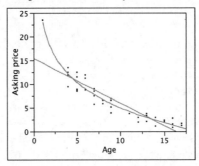

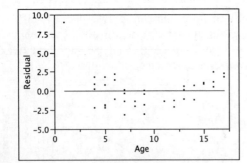
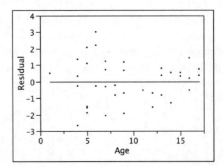

37. **Cellular Phones in the United States**
(a) The rapid expansion of this market suggests that we ought to expect something nonlinear -- exponential growth.
(b) It's certainly growing rapidly, perhaps more smoothly than we might have expected.
(c) The linear equation is not a good summary. The fitted equation
$$\text{Estimated Subscribers} = -1.96\text{e}+10 + 9{,}831{,}282.4 \, Date$$
suggests that the market is growing at about 10 million per year. That's in the right ballpark overall, but is much

Chapter 20 Curved Patterns

too fast for the early years and much too slow for the later years. The negative intercept is a reminder that we cannot extrapolate this equation back to the year zero.

(d) On the log scale, the trend in the data bends "the other way". An equation that's linear in the logs misses the trend as well. For example, the linear equation under-predicts the later years. This equation over-predicts the number of subscribers in later years.

(e) With the percentage changes plotted on the years since 1984, we can see the gradual slowing of the rate of growth. This rate of growth would be roughly constant for the log model to describe the curve. Since the rate slows, the log model misses the curvature.

(f) The estimated equation for the curve is
   Estimated Pct Growth = $1.1793435 + 148.0128 \cdot 1/(Date-1984)$
As the following scatterplot shows, this curve captures the slowing growth.

(g) For many, many years past 1984 (basically, infinitely many years), the rate of growth will eventually slow to 1.18 percent.

(h) Based on this fitted curve, we'd estimate the percentage growth in the next period (*Date* - 1984 = 23) to be
   Estimated Pct Growth = $1.1793435 + 148.0128/23 \approx 7.6\%$

Applying this to the last observation predicts the number of subscribers to be $219{,}420{,}457 \times 1.076 \approx 236{,}096{,}411$ – about 236 million.

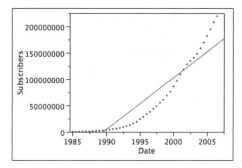
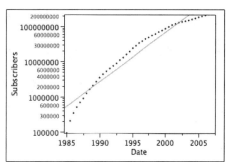

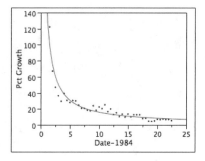

39. **Pet Foods, Revisited**
    (a) The content of the plots are the same; only the labels on the axes have changed. Natural logs (on the left) are larger than base 10 logs, by a constant factor (see part "d").

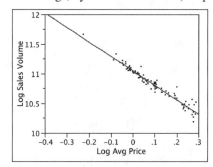
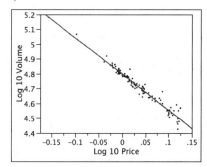

(b) *Estimated Log Sales Volume* = 11.050556 − 2.4420491 *Log Avg Price*
For base 10 logs b_1 is the same, but the intercept is smaller
Estimated Log 10 Volume = 4.7991954 − 2.4420491 *Log 10 Price*
(c) The r^2 of both fits is 0.9546, and the s_e of the log 10 equation is smaller (0.026254 versus 0.060453) by the same factor that distinguishes the intercepts, 0.060453/0.026254 = 2.30262.
(d) The log 10 and log e differ by a constant factor,
  $\log_e x = 2.30262 \log_{10} x$
The slope, this constant factor, is the $\log_e 10$. Hence, we can substitute one for another in equations, as long as we keep the factor 2.30262 in the right place.

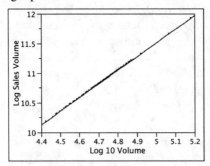

(e) Use any base that you'd like, so long as both x and y use the same base. The elasticity is the same.

Chapter 21 The Simple Regression Model

Mix and Match

1. e
3. b
5. d
7. l
9. k
11. g

True/False

13. False. The errors around the assumed linear pattern are normal, not the response.

15. True

17. False. To estimate the slope, we need variation in the explanatory variable.

19. True. Assuming the x's are comparable to those used in the existing data, then the SE would decrease to about $1/\sqrt{2} \approx 0.707$, or 70%, of its previous size.

21. True

Think About It

23. Yes. Set the slope β_1 to 0 and the intercept β_0 to μ.

25. (a) $b_0 \approx \$100$ and $b_1 \approx 2$. Estimate the intercept by where the line hits the y-axis. For the slope, notice that the line goes up to almost 500 when $x = 180$. That's a rise of nearly 400 for a change in x of 200.
    (b) Less than 1. In fact se(b_1) ≈ 0.2. Following the estimation procedure used to guess b_1 in (a), could you really be off by so much as 1?
    (c) Approximately 0.5. Actually, $r^2 = 0.60$. The correlation itself is larger than $\sqrt{0.6} = 0.77$.
    (d) $s_e = \$45.54$. A good guess from looking at the size of the prediction intervals would be $s_e \approx \$50$. The length of these intervals is about 4 times s_e.

27. The SRM denotes the response as y and the explanatory variable as x. It does not say that these have to be the original observed columns of data.

29. The averages should fall roughly along a line, with comparable variation around each. The averages will not be exactly on a line because of sampling variation: we observe averages during 12-week periods, not the whole population.

31. The slope and intercept ought to be about the same. The value of s_e would be much smaller since it measures the variation of averages rather than individual cases. r^2 would probably also be much larger. This figure and output show the actual results by converting the 320 diamonds into 21 averages. r^2 grows to 0.85 and $s_e = \$69$. The fitted equation is rather similar to that obtained for the individual diamonds. The equation for diamonds has intercept $b_0 = \$43$ with slope $b_1 = \$2670$/carat with $r^2 = 0.43$ and $s_e = \$169$.

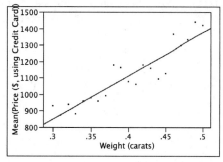

Estimated Mean(Price $) = 67.679761 + 2617.4834 Weight (carats)

33. β_1 remains the same, since it basically has no units (the units cancel). β_0 carries the units of the response, and so would be 100 times larger when using percentage changes rather than returns. r^2 remains the same, and s_e changes like the intercept.

35. (a) The estimated fit is $\hat{y} = 6.993459 + 0.5134397\,x$. Sampling variation explains why $\beta_1 \neq b_1$. β_1 is the population parameter, whereas b_1 is an estimate for one sample.
    (b) Yes, both intervals include the associated parameters. The interval for β_0 is $6.99 \pm 2(0.18)$ and for β_1, $0.513 \pm 2(0.0299)$. The interval for β_0 contains 7 and the interval for β_1 contains 0.5.
    (c) r^2 will stay about the same (we'd need to figure out the population correlation to be sure where this one is headed). s_e will get closer to $\sigma_\varepsilon = 1.5$, so it's likely to rise. Similarly β_0 and β_1 will get closer (most likely) to 7 and 0.5, respectively. The standard errors on both will be much, much smaller.

You Do It

37. **Diamond rings**
    The pattern is about as straight as they come (at least over this range of the sizes of diamonds), and the fit looks fine otherwise as well. The variances are similar and the residuals (in the normal quantile plot) are nearly normal. There might nonetheless be a lurking factor, but it's not evident in these views of the data.

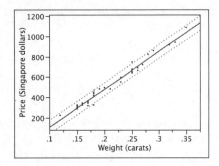
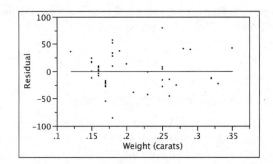
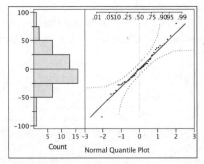

(a) The intercept is a large extrapolation. The confidence interval, even adjusting for this degree of extrapolation does not include 0. b_0 is about 15 $se(b_0)$ below zero (the *t*-statistic for the intercept), too far to happen by chance alone. In this case, the significantly negative intercept is a reminder that this equation is suited only to the narrow range of weights considered in this analysis. We'd expect a *positive* intercept, representing the fixed cost of the setting for the diamond (the gold band, for example).

(b) The prediction interval for the price of a ring with a ¼ carat diamond is
$(-259.6259 + 3721.0249 \times 0.25 = 670.630325)$ \$S $\pm t_{0.025,46}\,s_e$
$= (670.630325 - 2.013 \times 31.84052,\ 670.630325 + 2.013 \times 31.84052) = (606.5354,\ 734.7253)$

Because \$800 lies well above this interval, we'd say that this ring must have a very, very nice diamond to command such a large price. It's usually expensive for its weight. (Alternatively, just count the number of s_e's that separate \$800 from the predicted value, $(800 - 670.630325)/31.84052 = 4.0630515770$. That's a long way from the predicted value, farther than we'd expect by chance.)

r^2	0.978261
s_e	31.84052
n	48

Chapter 21 The Simple Regression Model   67

Term	Estimate	Std Error	t statistic	p-value
Intercept	−259.6259	17.31886	−14.99	<.0001
Weight (carats)	3721.0249	81.78588	45.50	<.0001

39. Download

These data meet the conditions of the SRM. The pattern is straight enough and the residuals seem simple enough. We don't know enough to suggest a lurking factor (such as time of day, perhaps). The variation might be increasing slightly with file size, but the effect is small. The residuals are nearly normal; the "stair–step" effect in the normal quantile plot suggests that the times have been rounded.

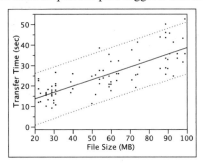
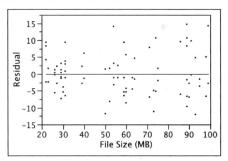

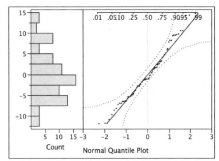

(a) Yes, we reject H_0: $\rho = 0$ because the slope differs significantly from zero. We reject H_0: $\beta_1 = 0$ because zero is not in the 95% CI, t-statistic > 2, and the p-value is less than 0.05.

(b) This is the confidence interval for the intercept ($t_{0.025,78} = 1.991$)

$7.27466 - 1.991 \times 1.71491, 7.27466 + 1.991 \times 1.71491 = 3.860274$ to 10.68905 sec

Since $se(b_0)$ rounds to 1.7 secs, so we should present this interval to one decimal place as 3.9 to 10.7 sec.

(c) The difference in times is

$7.2746633 + 0.3133071 (50) - (7.2746633 + 0.3133071 (50/2)) = 25\, b_1$

So, we need a CI for 25 times b_1. We can either take 25 times the endpoints of the CI for β_1, or multiply b_1 and its se by 25 and then form the CI. Since we've already got the CI in (c), we'll use that:

$25 \times 0.2582971, 25 \times .3683171 = 6.4574275$ to 9.2079275 sec

To set the rounding precision, we need the associated scale for this interval, namely $25\, se(b_1) = 25 \times 0.027505 = .687625$ which rounds to 0.07 sec. So, we round to tenths of a second and present the interval as 6.5 to 9.2 seconds saved.

r^2	0.624552
s_e	6.243347
n	80

Term	Estimate	Std Error	t statistic	p-value
Intercept	7.2746633	1.714906	4.24	<.0001
File Size (MB)	0.3133071	0.027505	11.39	<.0001

41. Seattle homes

The reason for the transformation to cost per square feet is that the variation increases as the homes become more expensive. The lack of constant variation about the fit is more evident in the plot of the residuals from this fit.

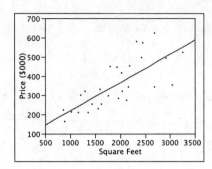
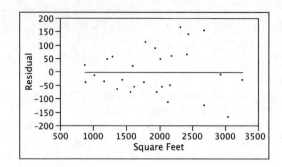

After transforming, the fit looks better, suited to the SRM.

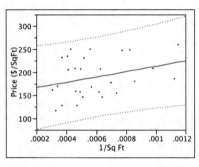
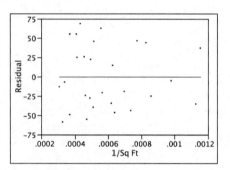

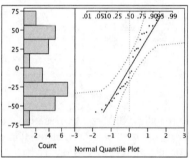

(a) The 95% CI for fixed costs is the interval for the slope ($t_{0.025,26} = 2.056$)
    57923.342 − 2.056 × 34515.8, 57923.342 + 2.056 × 34515.8 ≈ -$13,041 to $128,887
Fixed costs might be zero, but could be considerable.
(b) The confidence interval for variable costs is that for the intercept, here
    155.72096 − 2.056 × 21.80695, 155.72096 + 2.056 × 21.80695 ≈ 111 to 201 $/SqFt
You have to pay dearly for homes in this area.
(c) Because the fixed costs might be zero, you could arguably ignore the slope in this calculation. We'll include it, taking the estimated value as our best guess. The estimated cost per square foot is then
    155.72096 + 57923.342/3000 = 175.02874067 $/SqFt.
The approximate 95% prediction interval is wide, reflecting the weak fit. Because we're looking at the cost of one home, use ± 2 s_e to set the range.
    175.02874067 − 2 × 41.27091, 175.02874067 + 2 × 41.27091
        = 92.48692067, 257.57056067 $/SqFt ≈ 90 to 260 $/SqFt.
(d) Multiply the prior endpoints by 3000
    92.48692067 × 3000 , 257.57056067 × 3000 ≈ $280,000 to $770,000

r^2	0.097731
s_e	41.27091
n	28

Term	Estimate	Std Error	t Stat	p-value
Intercept	155.72096	21.80695	7.14	<.0001

Chapter 21 The Simple Regression Model

Term	Estimate	Std Error	t Stat	p-value
1/Sq Ft	57923.342	34515.8	1.68	0.1053

43. R&D expenses

The use of the log transformation for x and y spread out the data, pulling in the outliers and spreading apart the smaller companies. (Here, we've used natural logs; base 10 logs change the labeling on the axis but show the same content.) The fit appears linear in this example as well, but describes more of the data rather than just the extremes. The residuals have similar variances, but are skewed, with more negative than positive variation (they spread out more below than above the line). As a result, the residuals are not nearly normal. That said, the residuals are closer to normal than those in the original units of the data.

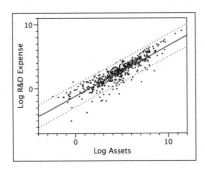
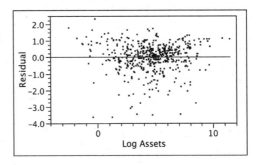
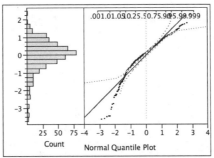

(a) The estimated slope is the elasticity. Each 1% increase in assets is associated in these data with a change of about 0.8% in the R&D spending ($t_{0.025,490} = 1.965$)

$0.790045 - 1.965 \times 0.01766, 0.790045 + 1.965 \times 0.01766 = 0.7553431$ to 0.8247469 %

or about 0.755 to 0.825 %

(b) No. The results are identical so long as we use the same log for transforming both x and y.

(c) The predicted log of R&D is for a firm with $1 billion in assets

$-1.184021 + 0.7900453 \times ln(1000) = 4.2734185916$

The 95% prediction interval (using $t_{0.025,200} = 1.972$ as a conservative estimate) for the log of R&D expenses is then $4.2734185916 - 1.972 \times 0.885719, 4.2734185916 + 1.972 \times 0.885719$

$= 2.526780724$ to 6.02005646

To move to the scale of dollars, we have to change these endpoints by taking the exponential of each:

$\exp(2.526780724), \exp(6.02005646) = 12.51315791$ to 411.601834

which we'd round to $12.5 to $412 million - a rather wide interval for a model that seems to look so predictive on the log scale.

r^2		0.80332
s_e		0.885719
n		492

Term	Estimate	Std Error	t Stat	p-value
Intercept	-1.184021	0.090258	-13.12	<.0001
Log Assets	0.7900453	0.01766	44.74	<.0001

45. **OECD**

The conditions for the SRM seem satisfied, though there are some problems. The pattern, such as it is, seems reasonably linear, so we'd say straight enough. There might be some trend away from the line for the smaller values, but with little data, it's hard to say. This might also represent a lack of similar variances. The problems continue into the residuals. With so few cases, these could be a sample from a normal population, but we'd bet against it. One can also think of many lurking factors, such as the effects of geographic or political connections. It's hard to think of countries that border one another as producing independent results.

In fact, 4 cases are from Eastern Europe (Hungary, Poland, Czech Republic, and Slovakia denoted by squares). The presence of such a strong lurking factor is perhaps more important in this example than the other visible flaws.

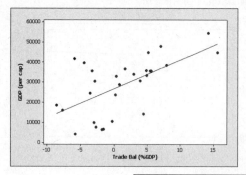
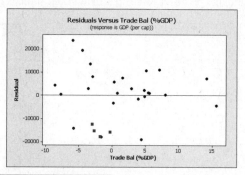
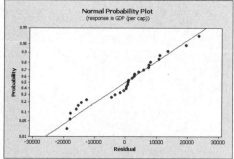

(a) Use the intercept from the model, so the CI is ($t_{0.025,27} = 2.051$)
26714.452 − 2.051 × 2145.280, 26714.452 + 2.051 × 2145.280 = $22,314.49 to $31,114.41.

(b) The slope indicates that on average, differences of 1% in trade balances are associated with a difference on average of 1440.510 − 2.051 × 362.497, 1440.510 + 2.051 × 362.497 = $697.03 to $2183.99

Her claim is equivalent to $2,000 per 1%, which lies inside this confidence interval. (Alternatively, note that (2000 − 1440.51)/362.497 ≈ 1.5 implies that the estimate is about 1.5 $se(b_1)$ above the estimated slope and thus inside the confidence interval.)

(c) The predicted per capital GDP is just the intercept; the slope drops out
$$26714.452 + 1440.510 \times 0 = 26{,}714.45$$
The 95% prediction interval for a country with balanced trade is then
$26714.452 - 2.051 \times 11335.782 \times \sqrt{1+1/29}$, $26714.452 - 2.051 \times 11335.782 \times \sqrt{1+1/29}$ = $3067.30 to $50,361.60 per person.

(d) The range in (c) is so much larger than the confidence interval because we're trying to predict what happens in one country, not what happens on average in the population. The intercept is like estimating the mean μ whereas the prediction interval is trying to predict one case. Hence the prediction interval has to allow for the country-to-country variation in setting the range.

r^2	0.369
s_e	11335.782
n	29

Term	Estimate	Std Error	t Ratio	Prob>|t|
Intercept	26714.452	2145.280	12.45	<.0001
Trade Bal (%GDP)	1440.510	362.497	3.97	<.0001

47. Promotion

These data seem well suited to the simple regression model. Although the association is weak, the pattern such as it is appears linear. These data meet the straight-enough condition. The residuals scatter randomly about the fitted model, with comparable variation. With so much variation unexplained, there are evidently lots of other explanations for the variation in share over time, but none is evident here. The residuals are also nearly normal. With a small sample, however, many samples appear nearly normal unless very far from bell-shaped.

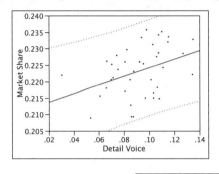
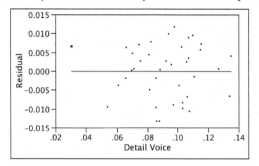
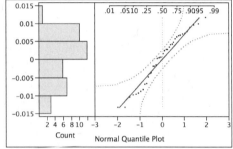

(a) Yes, though small, the association is statistically significant (zero is not in the 95% CI for the slope, the t-statistic for the slope is larger than 2, and the p-value for the slope is less than 0.05).

(b) Yes. The confidence interval for the intercept is ($t_{0.025,37}$ = 2.026)

$$0.211254 - 2.026 \times 0.004967, 0.211254 + 2.026 \times 0.004967 = 0.2012 \text{ to } 0.2213$$

or just above 20 to more than 22 percent of the market. (You could do a one-sided analysis, but this is so much easier!)

(c) No, not on the basis of this model. The promotion has not proven itself cost effective. The confidence interval for the slope

$$0.13005 - 2.026 \times 0.05260, 0.13005 + 2.026 \times 0.05260 = 0.02348 \text{ to } 0.2366$$

includes the 0.16 "break-even" coefficient. Increasing promotion *might* be effective. More likely, the promotion would not be cost effective. The estimated value indicates that on average the increase in promotion would not be effective because 0.13 < 1/6.

(d) The range for the intercept and predicted value would have to be multiplied by 100. The slope would remain the same. Other conclusions, those that involve statistical significance, are the same.

r^2	0.142129
s_e	0.007125
n	39

Term	Estimate	Std Error	t Ratio	Prob>|t|
Intercept	0.211254	0.004967	42.53	<.0001
Detail Voice	0.13005	0.05260	2.47	0.018

Chapter 22 Regression Diagnostics

Mix and Match

1. i or f
3. f or i
5. b
7. e
9. j

True/False

11. False. When the variance of the errors is not constant, the prediction intervals are likely to be too short in some cases (where the variance is large) and too long in others (where the variance is small).

13. True

15. True

17. False. Residuals look normally distributed when all of those omitted lurking factors combine, so their net affect - the sum of their effects - tends to have bell-shaped variation.

19. False. The decision to exclude data should take into account many things, particularly the substantive importance of the case. That unusual case might be the most interesting part of the data, telling you what you have omitted from the model.

21. False. Because it removes the trend, it's easier to see changes in the variation in the plot of the residuals on x.

Think About It

23. The data would most likely have unequal variation, with more variation among larger stores. Some large stores will do well (little competition, lots of people nearby) whereas others will not. Smaller locations have a smaller range of opportunities to do well, as well as a smaller down side. You might anticipate other problems as well.

25. The analyst was hasty because the analyst failed to realize that the problem with the regression is an evident lack of constant variation. These residuals should not be combined into one histogram. The variance is clearly smaller at the left than the right.

27. (a) The slope will become closer to zero.
    (b) The r^2 would change, but it is hard to say by how much. (In fact, it drops from 0.24 to 0.21.) Indeed, we must be very careful comparing r^2 for equations that do not describe the same variation. s_e would be smaller without this one, since it is the largest residual in the data.
    (c) Yes, this case is leveraged because it is near the right-hand side of the plot.

29. (a) The slope will increase, moving to near zero.
    (b) R^2 will decrease, whereas s_e will stay about the same or be slightly smaller. r^2 decreases because most of the variation that is explained is the distinction of this outlier from the rest of the cases (r^2 drops from about 30% to near 10%.)
    (c) This case is leveraged because it lies far below the other cases.

31. Answers will vary, but you can think of other macroeconomic factors that were also changing over this time period, such as trends in the stock market, interest rates, inflation, etc.

33. No. The Durbin-Watson statistic tests the assumption of independence. If we do not reject this hypothesis, it may still be false. We just failed to reject it. We have not proven that independence is true. Statistical tests never prove H_0.

You Do It

35. Diamond rings

The price of the Hope Diamond comes to S$ 56 million.
(a) With this point added, the scaling on the plot is such that you can see only 2 points: one for the Hope Diamond, and one for the other 48 points.
(b) The fitted line essentially goes through these 2 points, as summarized below. The slope becomes much, much steeper. The intercept becomes even more negative.
Without Hope Diamond (green, nearly horizontal in figure)
   Estimated Price (Singapore dollars) = -260 + 3,721 Weight (carats)
With Hope Diamond (red)
   Estimated Price (Singapore dollars) = -251,696 + 1,235,667 Weight (carats)
(c) The value of R^2 grows from 0.978 to 0.999925 and s_e gets huge, swelling from S$32 to S$69,962. We should not directly compare R^2 because we've changed the response. It's grown so large because most of the variation in the new response is the difference between rings with little diamonds and this huge stone. s_e is larger because even a small error in fitting the Hope Diamond is large when compared to the costs of the other rings. The model is not even close to them now. (See the scatterplot on the right that shows the fit of the new equation to the small rings; the fit is the vertical line in the figure!)
(d) The point for the Hope Diamond is incredibly leveraged, with a value of x that is about 100 times heavier than any other in the data. The least squares regression has to fit this outlier, no matter what it does to the fit the other data.

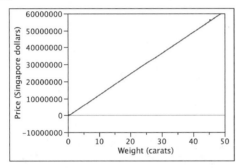
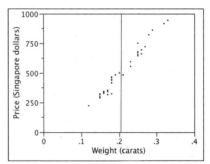

Term	Estimate	Std Error	t Ratio	Prob>|t|
Intercept	-251696.1	10148.5	-24.80	<.0001
Weight (carats)	1235667.3	1559.812	792.19	<.0001

37. Download

(a) Neither plot suggests a problem. The fitted equation is
   Estimated Transfer Time (sec) = 7.2746633 + 0.3133071 *File Size* (MB)
The residual plot versus both the explanatory variable and the time order seem fine at first glance. Both plots are shown below.
(b) The Durbin-Watson D statistic is $D = 2.67$. For a sequence of this length, this is statistically significantly different from 2. Our software computes the p-value at 0.003.
(c) In this example, the pattern is one that we have not seen. Rather than showing the meandering pattern, these flip sign. The residuals basically have the pattern positive/negative/positive/negative with alternating sign.

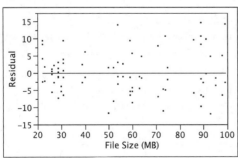
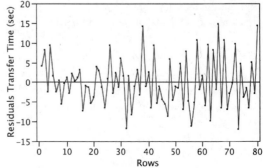

39. **Seattle homes**
(a) The two fits are shown together in the plot below. The near horizontal line includes the outlier. The line with steeper slope does not.
(b) Use the model without the outlier as a basis for setting the size of confidence intervals. The estimates with the outlier are not very close to those obtained without this home, but the slope nonetheless falls within the range of uncertainty indicted by the confidence intervals (we only have a small sample, so these intervals are wide). For the slope, the gap between the estimates is
$$(5175.4905 - 57923.342)/34515.8 \approx -1.5$$
and for the intercept, the gap is smaller in absolute size, but considerable on the standard error scale:
$$(201.01784 - 155.72096)/21.80695 \approx 2.1$$
(c) The intercept represents variable costs (estimated to be $156 per square foot without the outlier). This estimate is more affected by the outlier than the slope. While the slope changes by more in absolute terms, the change is within the realms of plausibility. The intercept lies outside the confidence interval if the outlier is included.
(d) Yes, the lot for this home is more than 3 times larger than any other. You're getting a lot more land with this home than the others, helping to explain why this home costs 3 times as much as others of the same number of square feet of house. With the lot size taken into account, the cost of this property seems in line with that of the home in row 23 (which costs $575,000 for 2452 square feet of home on a lot with 248,000 square feet).

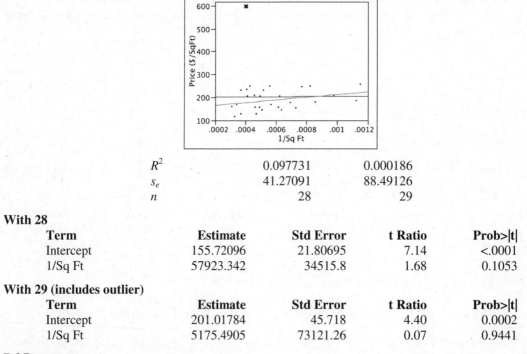

R^2	0.097731	0.000186
s_e	41.27091	88.49126
n	28	29

With 28

Term	Estimate	Std Error	t Ratio	Prob>|t|
Intercept	155.72096	21.80695	7.14	<.0001
1/Sq Ft	57923.342	34515.8	1.68	0.1053

With 29 (includes outlier)

Term	Estimate	Std Error	t Ratio	Prob>|t|
Intercept	201.01784	45.718	4.40	0.0002
1/Sq Ft	5175.4905	73121.26	0.07	0.9441

41. **R&D expenses**
(a) The plot has an odd "flat-top" appearance, with the variation above the fitted line being smaller, more compact than that below the line. Notice the scale in the residual plot. Negative deviations seem much more spread out than positive deviations.
(b) ½. The normal distribution is symmetric.
(c) We counted 24 companies whose values lie outside the indicated prediction intervals. Of these, only 2 are positive. We'd expect half, or 12. The SD of a binomial with $n = 26$ and $p = \frac{1}{2}$ is $\sqrt{np(1-p)} = \sqrt{24/4} \approx 2.45$. That means the observed count of 2 lies $(2 - 12)/2.45 = -4.1$ SD's below the mean. The central limit theorem (applied to the binomial) tells us that this is rather unusual. Seems that indeed the errors are not nearly normal. (The quantile plot confirms this impression, but its nice to know some alternatives if you can't do the quantile plot easily.)

Chapter 22 Regression Diagnostics

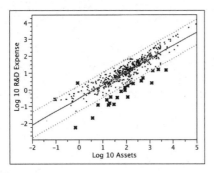
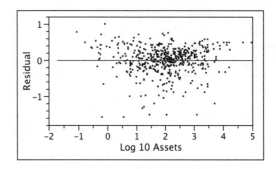

43. **OECD**
(a) Visually, the fit does not change by very much, as shown in the plots below. The two fitted questions produce very similar fits to the data. More precisely, we can use the confidence interval for the fit based on all of the data. The fitted equations using all of the data and then without Luxembourg are
All 30 countries
  Estimated GDP (per cap) = 26,804 + 1,617 Trade Bal (%GDP)
Without Luxembourg
  Estimated GDP (per cap) = 26,714 + 1,441 Trade Bal (%GDP)
The confidence interval for the slope using all of the data is
  $1617.47 - 2 \times 303.86, 1617.47 + 2 \times 303.86 = 1009.75$ to 2225.19
The slope without Luxembourg is well within the confidence interval.

(b) These summary statistics change quite a bit. As always, we have to be careful comparing the values of R^2 since we have changed the response by removing a case.

	R^2	s_e
All	0.503	11,298
Without Luxembourg	0.369	11,336

The change in R^2 is so much larger because Luxembourg is also the largest value on the response. When we remove it, we remove a large contributor to the variation on the y scale, variation that we had been explaining. s_e changes relatively little since the fit remains the same and the residual at Luxembourg was fairly typical of those at other points.

(c) No. The regression does not take into account the sizes of the countries. All are equally weighted. That's a problem in the sense that data for a small country might be more variable from year to year than that for a larger country. Think of the analogy to averages: averages of larger samples are more stable than averages of smaller samples.

45. **Promotion**
(a) Week 6 had unusually (and suddenly) low levels of detailing. Voice had been a steady 10% before suddenly falling off. Sales remained steady.

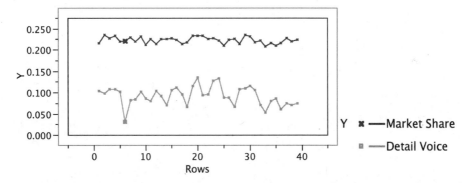

(b) Relative to the fit of the model without the outlier, the fit with the outlier gives similar estimates. The change with the outlier in the intercept is
  $(0.2111775 - 0.2082504)/0.005646 = .5184378320$ (b_0 is larger with the outlier)
and for the slope

$(0.1313929 - 0.1607747)/0.059047 = -.497600216$ (b_1 is smaller)

Both changes are on the order of about ½ of a standard error, well within the range of plausibility suggested by the confidence intervals from the model without the outlier.

(c) Week 6 is highly leveraged, so it increases the variation in the explanatory variable. Without this case, we have less variation in x and hence get a larger standard error for the slope, even though s_e is smaller. We also have a smaller n without the outlier.

(d) The Durbin-Watson statistic $D = 2.02$ and the timeplot of the residuals shows no pattern. There's no evidence of a lurking factor over time.

	With	Without
R^2	0.14467	0.170771
s_e	0.007102	0.007086
n	39	38

With all 39

Term	Estimate	Std Error	t Ratio	Prob>|t|
Intercept	0.2111775	0.004964	42.54	<.0001
Detail Voice	0.1313929	0.052523	2.50	0.0169

Without the outlier

Term	Estimate	Std Error	t Ratio	Prob>|t|
Intercept	0.2082504	0.005646	36.88	<.0001
Detail Voice	0.1607747	0.059047	2.72	0.0099

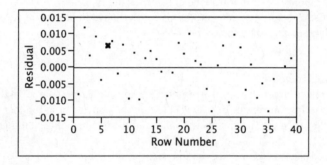

Chapter 23 Multiple Regression

Mix and Match

1. h or e

3. c

5. f

7. a

9. b

True/False

11. True

13. False. It's called a marginal slope because it includes the effects of other explanatory variables.

15. False. It might be smaller, but it does not have to be smaller. It depends on the size and sign of any indirect effects.

17. True

19. False. We should only conclude that at least some deviation from this hypothesis occurs. It may not be the case that both are different from zero. Perhaps only one of them differs from zero.

21. False. Its primary use is checking the similar variances condition.

Think About It

23. Most likely we have some collinearity. Busy areas attract a lot of fast food outlets because sales are high (positive correlation). Among densely populated areas, however, the number of competitors reduces sales of a store (negative partial slope). You'd like to have the densely populated area to yourself. The more competitors that are around, the lower your sales for a give population density.

25. (a) Estimated Salary = b_0 + 5 Age + 2 Test Score
    (b) The indirect effect is 10 $M/Point = 2 years/point × 5 $M/year, larger than the direct effect.
    (c) The marginal effect is the direct plus indirect effect, or 10 + 2 = 12 $M/point.
    (d) You're not going to be much older, so we need the partial effect. Raising the test score by 5 points nets $10,000 annually. It's probably worth it if you're going to stay with the company long enough to earn it back.

27. (a) The correlation of something with itself is 1.
    (b) You cannot, not without knowing the variance of x_1 and y.
    (c) The partial and marginal slopes will be the same because the two explanatory variables are evidently uncorrelated. There can be no indirect effect.

29. The order of the variables in the correlation matrix is Z, X, T, Y. The easiest way is to start with the large positive correlation between Z and X and work from there.

31. (a) The fitted value is
    87 + 0.3 × 250 + 1.5 × 100 = 312, or $312,000 revenue per month
    87 + 0.3 × 200 + 1.5 × 75 = 259.5, or $259,500 revenue per month
    Expand to the second location.
    (b) The intercept, $87,000, resembles a fixed cost. The intercept estimates fixed revenue that is present regardless of the distance to the destination or the population. Perhaps it's money earned from air freight or other services provided by the airline. Without a confidence interval, we cannot be sure if the value is really far from zero. It might be a large extrapolation.
    (c) Among comparably populated cities, flights to those that are 100 miles farther away produce 0.3 × 100 = $30,000 more revenue per month, on average.

(d) If we compare revenue from flights to cities that are equally distant from the hub, average monthly revenue to larger cities is higher by about $1.5 per person.

33. (a), (b) The filled in table is

	Estimate	**SE**	***t*-statistic**	***p*-value**
Intercept	87.3543	55.0459	1.5869	≈ 0.10
Distance	0.3428	0.0925	3.7060	< 0.01
Population	1.4789	0.2515	5.8803	< 0.01

(c) Yes, the *t*-statistic for *Distance* is larger than 2 in absolute size.

(d) Based on the fit of this model, the confidence interval for 10 times the slope for population is
    14.789 − 2× 2.515, 14.789 + 2× 2.515 = 9.759 to 19.819 thousand dollars
The relevant se rounds to 2.5, so we should keep 1 decimal place and give the interval as 9.7 to 19.8 thousand dollars ($9,700 to $19,800).

35. (a) Yes. The overall F-statistic is $(0.74/0.26) \times ((37 − 1 − 2)/2) \approx 48.4 \gg 4$.

(b) The standard deviation of the residuals around the fit is $32,700. Given that the conditions of the model check out, we ought to be able to predict monthly revenue to within about $65,000 with 95% confidence.

You Do It

37. **Gold Chains**

(a) The plots show the discrete properties of the data: we only have several fixed lengths and widths. Width is very highly related to price. The two *x*s are not very correlated. The plots look straight enough (particularly that for width).

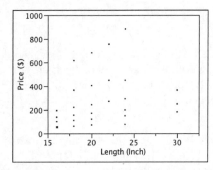
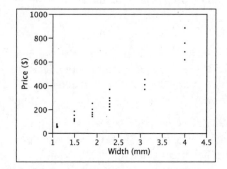
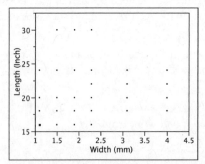

(b) The largest correlation (0.95) is between price and width. Evidently width tells you more about how much gold than the length.

	Price ($)	Length (Inch)	Width (mm)
Price ($)	1.0000	0.1998	0.9544
Length (Inch)	0.1998	1.0000	0.0355
Width (mm)	0.9544	0.0355	1.0000

(c) The fit of this model has $R^2 = 0.94$ and $s_e = \$57$ with these coefficients…

Term	Estimate	Std Error	t Ratio	Prob>|t|
Intercept	-405.635	62.11863	-6.53	<.0001
Length (Inch)	8.8838083	2.654034	3.35	0.0026

Chapter 23 Multiple Regression

Term	Estimate	Std Error	t Ratio	Prob>|t|
Width (mm)	222.48894	11.64679	19.10	<.0001

(d) First, the overall fit of the model is not straight enough; there's a trend in the residuals. Second, the model is missing an obviously important variable: the amount of gold in the chain.

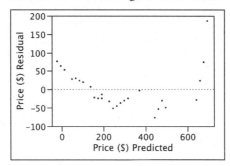

(e) We formed the "volume" of the chain as the length (in mm) times the width^2. This in a way gets at the amount of gold in the chain, though not perfectly. The residuals have some pattern left, but there's not the clear trend as before, and now we can identify some outliers (a bargain and an expensive chain) that were hidden.

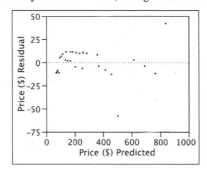

(f) Here's the fit for the improved model. With the added volume, the other two explanatory variables, particularly the length, lose importance. The model looks much straighter with a much smaller s_e near $17. There's still a problem in the residuals, but they are much smaller. Our proxy for gold isn't perfect for the heavier chains.

R^2	0.994674	
s_e	17.0672	

Term	Estimate	Std Error	t Ratio	Prob>|t|
Intercept	55.118884	34.43198	1.60	0.1225
Length (inch)	0.0451975	0.971144	0.05	0.9633
Width (mm)	-30.59663	16.27885	-1.88	0.0724
Volume (cu mm)	0.0930388	0.005845	15.92	<.0001

39. **Download**
(a) The file sizes increased steadily over the day, meaning that these two explanatory variables are closely associated. The scatterplots of transfer time on file size and time of day seem reasonably linear, though their may be some bending in the plot of transfer time on the time of day.

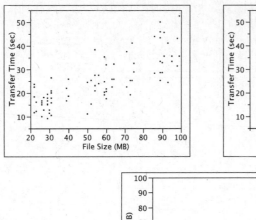
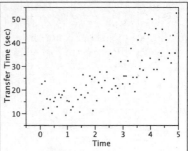
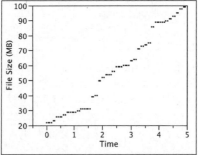

(b) The marginal and partial slopes for the file size will be very different. We will not easily be able to separate their influence from one another. The file size and time of day are virtually redundant, so the indirect effect of file size will be very large.

(c) The multiple regression is

R^2		0.624569
s_e		6.283617

Term	Estimate	Std Error	t Ratio	Prob>|t|
Intercept	7.1388209	2.885703	2.47	0.0156
File Size (MB)	0.3237435	0.179818	1.80	0.0757
Time (hours since 8 am)	-0.185726	3.16189	-0.06	0.9533

(d) Somewhat, but not completely. The residual plot suggests slightly more variation for larger file sizes. The effect is fairly subtle and is also evident in a time plot of the residuals. There is also a slight negative dependence over time, with the residuals oscillating back in forth from positive to negative. Again, the effect is not too strong (albeit significant by the Durbin-Watson test, $D = 2.67$). The residuals appear nearly normal with no evidence of bending patterns.

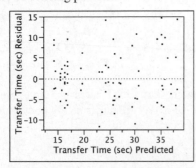
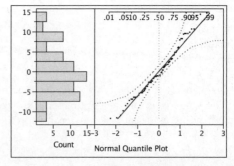

(e) No. The overall F-statistic is approximately $F = (0.624/(1 - 0.624)) \times (77/2) \approx 64$ and is very significant (being much larger than 4). On the other hand, the t-statistics as seen in the tabular summary are both less than 2. Thus, we can reject $H_0: \beta_1 = \beta_2 = 0$, but cannot reject either $H_0: \beta_1 = 0$ or $H_0: \beta_2 = 0$.

(f) The key difference is the increase in the SE of the slope. The confidence interval for the partial slope for file size from the multiple regression is $0.3237435 - 2 \times 0.179818$ to $0.3237435 + 2 \times 0.179818$, or about -.04 to 0.68 seconds per MB - a huge range that includes zero. The marginal slope is $0.3133 - 2 \times 0.0275$ to $0.3133 + 2 \times 0.0275$, or about .2583 to .3683 seconds per MB. The estimates (slopes) are about the same, but the range in

Chapter 23 Multiple Regression

the multiple regression is much larger.

(g) The direct effect of file size (from the multiple regression) is indirect effect of file size is 0.32 sec/MB. The indirect effect (from the simple regressions) is

(0.0562 hours since 8am/MB) × (-0.186 sec/hour after 8am) = -.0104532 sec/MB

is very small. The path diagram only tells you about the difference between the indirect and direct effect (slope in the simple and multiple regression), not the change in the standard errors.

41. **Home prices**

(a) Some of the homes are large and expensive, making these leveraged outliers. The relationships appear linear. The two explanatory variables are related, as you would expect.

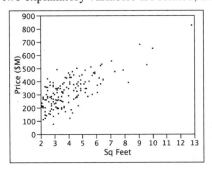 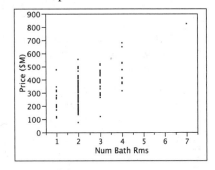

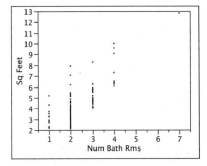

(b)

R^2		0.533512
s_e		81.03068

Term	Estimate	Std Error	t Ratio	Prob>|t|
Intercept	107.41869	19.59055	5.48	<.0001
Sq Feet	45.16066	5.78193	7.81	<.0001
Num Bath Rms	14.793861	11.74715	1.26	0.2099

(c) There's no sign of the usual changing variation. This looks to meet the usual assumptions. The concern remains the presence of the leveraged outlier. The residuals are nearly normal.

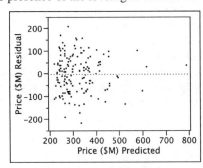 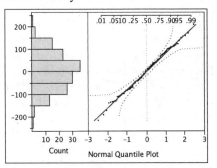

(d) Yes. The overall F-statistic is $F = (0.5335/(1 - 0.5335)) \times (150 - 1 - 2)/2 \approx 84$ which is much larger than 4 needed to assure statistical significance.

(e) The confidence interval for the marginal slope is
  $82.3267 - 2 \times 9.4291, 82.3267 + 2 \times 9.4291 = 63.4685$ to 101.1849
or about 63 to 101 thousand dollars per bathroom. For the partial slope, the CI is
  $14.7939 - 2 \times 11.7472, 14.7939 + 2 \times 11.7472 = -8.7005$ to 38.2883
or about -9 to 38 thousand dollars per bathroom. The range of the intervals is comparable, but the estimates are rather different. The estimates change because of the correlation between the two explanatory variables (evident in (a)) which implies a large indirect effect.

(f) She's unlikely to recover the value of the conversion from the sale price. The value of converting space (the partial slope; the conversion to a bathroom does not increase the size of the home) is from -9 to 38 thousand, and her cost of 40 thousand lies outside this range. She shouldn't do it.

43. **R&D expenses**
(a) The scatterplots (all on log scales) show strongly linear trends, but between y and the explanatory variables as well as between the explanatory variables

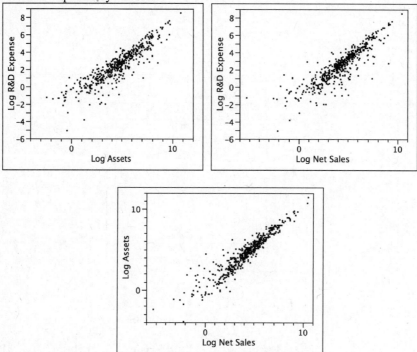

(b)

R^2	0.80991
s_e	0.869808

Term	Estimate	Std Error	t Ratio	Prob>|t|
Intercept	-1.203173	0.089859	-13.39	<.0001
Log Assets	0.5831633	0.052146	11.18	<.0001
Log Net Sales	0.2284876	0.053194	4.30	<.0001

(c) The residuals are skewed, even on the log scale. The range below zero is more extreme than the range above. That is, the variation of negative residuals is larger than the variation of positive residuals. As a result, the data are not nearly normal. The model would not be suitable for prediction (i.e., 95% prediction intervals would not have the right coverage). The central limit theorem suggests inferences about slopes are okay, but not for predicting individual companies.

Chapter 23 Multiple Regression

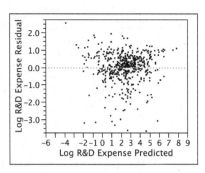
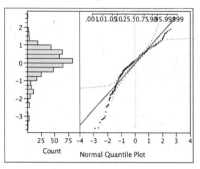

(d) Yes, because the t-statistic (4.3) indicates that this slope is significantly different from zero. Hence, the addition of this explanatory variable significantly increases R^2.

(e) The partial elasticity of R&D expenses with respect to net sales is

$0.2284876 - 2 \times 0.053194$, $0.2284876 + 2 \times 0.053194 = .1220996$ to $.3348756$

or about (to presentation precision) 0.12 to 0.33. Among companies of equal assets, R&D spending averages between 0.12 to 0.33 percent higher among those with 1% higher net sales.

(f) Yes, it's considerably smaller. The marginal elasticity is 0.79 ± 0.04, so the confidence intervals for the estimates do not even overlap. The simple explanation for the difference is that the partial elasticity estimates the effect of percentage differences in net sales among companies with equal assets. The marginal elasticity includes the indirect effect: the marginal elasticity includes the benefit of having more assets (which itself has positive partial elasticity).

45. **OECD**

(a) The scatterplots show a very strong association between y and the second predictor. This second variable appears more associated with the GDP, as well as having a more linear relation. The scatterplots seem reasonably linear.

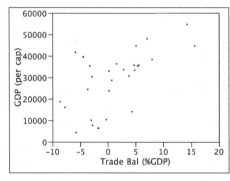
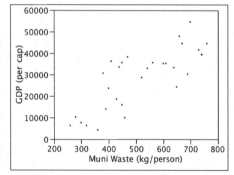

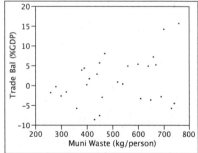

(b) The two xs are correlated ($r \approx 0.3$). The slope for the trade balance will change because of the presence of indirect effects.

(c) The estimated model is

$$R^2 \qquad 0.772618$$
$$s_e \qquad 6934.623$$

Term	Estimate	Std Error	t Ratio	Prob>|t|
Intercept	-4622.225	4796.003	-0.96	0.3440
Trade Bal (%GDP)	959.60593	232.7805	4.12	0.0003
Muni Waste (kg/person)	62.184369	9.153925	6.79	<.0001

(d) Yes. For example, the residuals have similar variances (left) and are nearly normal (right). Of course, with only 29 cases, we cannot be very sure and we may have missed a subtle problem.

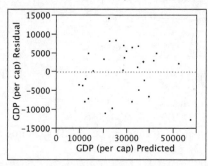
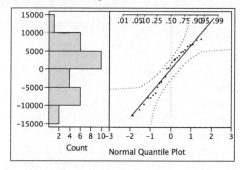

(e) The direct path from trade balance to y has coefficient 960 and the path from waste to y has coefficient 62. The path from trade balance to muni waste has slope from the fit

Estimated Muni Waste (kg/person) = 503.93174 + 7.7335205 Trade Bal (%GDP)

Similarly the path from municipal waste to trade balance has slope

Estimated Trade Bal (%GDP) = -4.990754 + 0.0119591 Muni Waste (kg/person)

The indirect effect for trade balance is thus

$7.7335205 \times 62.184369 \approx 481$

As a check the sum of the direct and indirect effects are

$960 + 481 = 1441$

which is the marginal slope for the trade balance. Because the indirect effect is positive, the marginal slope is larger than the partial slope. On average, countries with larger exports have more consumption (producing more trash), and this consumption contributes to GDP.

(f) The 95% confidence interval for the slope for municipal waste is

$62.1843 - 2 \times 9.1539, 62.1843 + 2 \times 9.1539 = \$43.8765$ to $\$80.4921$

more GDP per kilogram of waste. This would be rounded to $44 to $80. The interval does not include zero, so that β_2 is not zero. This does not mean countries should produce more waste. Rather, it means that at a given trade balance, countries with more waste per person have larger GDP per person. The model is not causal.

47. Promotion

(a) The scatterplots are vaguely linear, with weak associations between the two predictors and the response. The largest correlation is between the two explanatory variables, so marginal and partial slopes will likely differ.

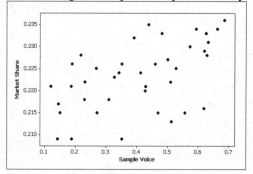

Chapter 23 Multiple Regression

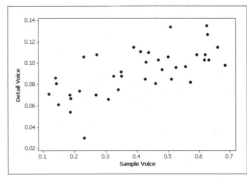

(b) The estimated model is

R^2		0.2794
s_e		0.006618

Term	Estimate	Std Error	t Ratio	Prob>|t|
Intercept	0.2128209	0.004652	45.74	<.0001
Detail Voice	0.0166246	0.065259	0.25	0.8004
Sample Voice	0.0219316	0.008364	2.62	0.0127

(c) The residuals look fine, though rather variable (*i.e.*, the model does not explain much variation.) The DW does not find a pattern over time ($D = 2.04$). The residuals are also nearly normal.

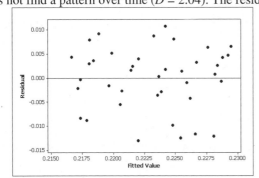
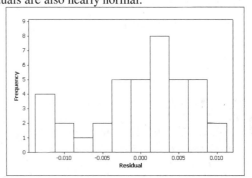

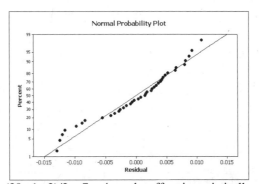

(d) Yes. $F = (0.28/(1 - 0.28)) \times (39 - 1 - 2)/2 \approx 7 > 4$, so the effect is statistically significant.
(e) No. The partial effect for detailing is not significantly different from zero.
(f) No. The model is not causal. The partial slope for detailing is not significantly different from zero (*i.e.*, zero is in the 95% confidence interval), but this does not mean detailing has no effect. It only means, as in the statement of the question in part (e), that at a given level of sample share, periods with a higher share of detailing have not shown gains in market share. Since detailing and sampling tend to come together, it is hard to separate the two. Perhaps the best advice would be to do some experiments.

Chapter 24 Building Regression Models

Mix and Match

1. f
3. i
5. j
7. b
9. c

True/False

11. True

13. True. In order for the model to obtain a large F-statistic, something has to be going on.

15. False. It's usually true, but not in every case. Suppose there are $n = 4$ observations. Then F = (0.8/0.2) × (1/2) = 2 which is not going to be statistically significant. You need to know both n and the number of predictors in the model.

17. True

19. False. You only need the values of the xs, not y, to compute the VIF.

21. False. This is one of many possible approaches. The best for any situation depends on the context of the problem. Often, the best approach is to (a) describe the nature of the collinearity and make sense of how the marginal and partial slopes differ (as in the 4M example) or (b) combine the redundant predictors.

Think About It

23. (a) In February, the market as a whole had positive returns, but the S&P was negative; the whole market went up, but the S&P did not. The opposite happened in the next month. The market as a whole rose about 5% above the risk-free rate, but the S&P shot up almost twice as much.
(b) Very much so. Relative to the pattern elsewhere, these months are rather different combinations of the two explanatory variables.
(c) Keep them. These points weaken the correlation between the two explanatory variables, supplying unique variation to the fitting. (The correlation between the two variables is 0.973 with them, and 0.982 without.)

25. The t-statistic for testing β_0 remains the same. The labels on the scatterplot of y on x change by a factor of 100, but the relationship stays the same. The value of β_0 would be 100 times smaller, but then its standard error would also be 100 times smaller. Hence, the ratio of these two, the t-statistic, would stay the same.

27. (a) The data occupy a grid with 2 points at every combination of income and age. For example, the average age is the same for every income and vice versa. If you were to view a scatterplot of age on income, you'd see a square grid of dots.
(b) Yes, exactly because *Age* and *Income* are uncorrelated (correlation is zero).
(c) You'd expect the slope for *Age* to be negative. By the sampling design, we've focused on the effect of *Age* among people with comparable incomes, and hence gotten at the partial slope.

29. (a) Not as designed. The two predictors will be perfectly correlated and cannot be separated. This is regression's version of confounding in the two-sample t-test.
(b) The analysis will certainly be simpler if the two factors (temperature and amount of additives) are varied independently of each other, perhaps over a range of commonly used settings. (A special group of techniques known as the "design of experiments" studies the creation of plans for varying several factors at once to arrive at an optimal combination with the least testing.)

Chapter 24 Building Regression Models

You Do It

31. **Gold Chains**
    (a) We can use both as explanatory variables, even though width is used to compute the volume. The two are not perfectly correlated since width gets squared and then multiplied by the length. The correlation is 0.966.
    (b) Yes, both improve the fit. The output below shows the estimated multiple regression. The model is statistically significant, with both variables significant (both t-statistics are larger than 2 in absolute size).
    (c) The correlation between the two variables is 0.966, so the VIF = $1/(1 - 0.966^2) \approx 15$. Collinearity increases the se of these variables by about $\sqrt{15} \approx 4$ times larger.
    (d) For chains of a given width, the retailer is charging about 12 cents per additional mm^3.
    (e) Collinearity; the negative slope for width indicates that our estimate of the amount of gold provided by the volume variable is off for the wider chains. There's more open space in the bigger chains, and hence less gold, so the slope for this predictor in the multiple regression is negative.

R^2		0.994674
s_e		16.72312
n		28

Term	Estimate	Std Error	t Ratio	Prob>|t|
Intercept	56.521547	16.31446	3.46	0.0019
Width (mm)	−31.01496	13.29882	−2.33	0.0280
Volume (cu mm)	0.1186587	0.005982	19.83	<.0001

33. **Download**
    (a) The overall F-statistic is roughly $(0.625/0.375) \times (77/2) \approx 64$, which is quite statistically significant.
    (b) No, because the absolute value of neither t-statistic is larger than 2.
    (c) The correlation between the predictors is 0.988, so the VIF is approximately $1/(1 - 0.988^2) \approx 42$. Collinearity has increased the standard errors of the estimated slopes by about $\sqrt{42} \approx 6.5$ times.
    (d) Yes, the two predictors are highly redundant because the transferred files got steadily larger during the testing. It's not possible to separate the time of day from the size of the file.
    (e) Yes, they could have randomly chosen the file to send at various times of day, rather than steadily increasing the size of the file as the testing went on.

R^2		0.624569
s_e		6.283617
n		80

Term	Estimate	Std Error	t Ratio	Prob>|t|
Intercept	7.1388209	2.885703	2.47	0.0156
File Size (MB)	0.3237435	0.179818	1.80	0.0757
Hours past 8	−0.185726	3.16189	−0.06	0.9533

35. **Home prices**
    (a) Yes. After all, each bathroom makes up some of the square feet in a home. Also, larger homes designed for more people also tend to have more bathrooms (though these days more and more homes have several bathrooms regardless of size).
    (b) The VIF for the predictors is about $1/(1 - 0.736^2) \approx 2.2$. Consequently, the estimated standard error for the slope of Bathrooms would have been smaller by a factor of $\sqrt{2.2} \approx 1.5$, making its t-statistic larger by this same factor. That is, the t-statistic for Bathrooms would have been about 50% larger, or $1.26 \times 1.5 = 1.89$. This is not quite enough to be statistically significant, but much closer.
    (c) The estimated slope in the regression of residuals is the *same* as that in the multiple regression. Even the estimated standard error is almost identical. The scatterplot of the two sets of residuals and fit of the simple regression follows the summary of the multiple regression.
    (d) The changes are twofold. First, there's less variation in the predictor. That's the unpleasant effect of collinearity. Second, there's less association between the unique variation that remains in these two sets of

residuals. We can see a trend marginally, but only a very weak pattern after removing the effects of size. It seems that once you know the size of a home, counting bathrooms doesn't tell you much more.

R^2		0.533512
s_e		81.03068
n		150

Term	Estimate	Std Error	t Ratio	Prob>|t|
Intercept	107.41869	19.59055	5.48	<.0001
Sq Feet (000)	45.16066	5.78193	7.81	<.0001
Bathrooms	14.793861	11.74715	1.26	0.2099

Term	Estimate	Std Error
Intercept	−1.2e−14	6.593737
Residuals Bathrooms	14.793861	11.7074

37. **R&D expenses**
(a) Yes, large corporations tend to have large values for both assets and sales, so these will be correlated.
(b) The correlation is about the same on either scale. On a log scale, the correlation is about 0.94. In dollars, the correlation is slightly larger, about 0.95
(c) In dollars, the correlation is heavily dominated by the two large outliers (Intel and Microsoft). The correlation is a better summary of the association after converting to logs. On a log scale, the data have a more reasonable linear, elliptical shape that is better summarized by the correlation.
(d) The VIF is about $1/(1 - 0.94^2) \approx 8.6$. The standard errors of the estimated slopes are $\sqrt{8.6} \approx 2.9$, or almost 3 times larger than if we had companies for which these two were uncorrelated. Even so, the variables are highly statistically significant.
(e) The estimated slopes are the same, with just a very slight difference in the estimates of the standard error.
(f) The range of *Log Assets* is reduced because of the collinearity; there's quite a bit less variation in the residuals of log assets than the log of assets themselves (a range of −3 to 11 vs. −3 to 4). The slope itself seems well preserved, and just a bit smaller. Collinearity increases the s_e but leaves the estimate.

R^2		0.80991
s_e		0.869808
n		489

Term	Estimate	Std Error	t Ratio	Prob>|t|
Intercept	−1.203173	0.089859	−13.39	<.0001
Log Assets	0.5831633	0.052146	11.18	<.0001
Log Net Sales	0.2284876	0.053194	4.30	<.0001

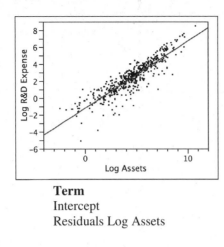
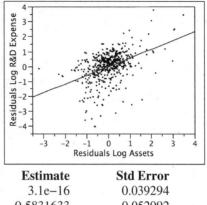

Term	Estimate	Std Error
Intercept	3.1e−16	0.039294
Residuals Log Assets	0.5831633	0.052092

39. **OECD**

    (a) In her model, there might be a reason for collinearity, but we couldn't think of one. Had the data been formulated as national totals, then most of the variables would be indirectly measuring the size of the economies and so be correlated.

    (b) Luxembourg is leveraged, with a relatively high trade balance.

    (c) No. The correlation between the two explanatory variables is only 0.34, so the VIF = $1/(1 - 0.34^2) \approx 1.13$. Collinearity increases the standard error slightly (by about $\sqrt{1.13} = 1.06$), but not much.

    (d) The partial slope matches the slope in the partial regression plot.

    (e) Luxembourg is even more highly leveraged in the regression analysis for the partial slope (right). Without Luxembourg, this view suggests that the partial slope would decrease, and its standard error would increase (because there would be less unique variation in this variable without Luxembourg). The output below shows this in fact happens.

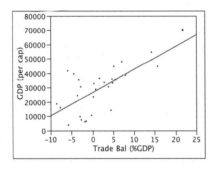
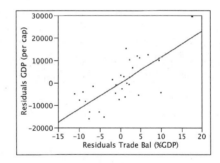

Term	Estimate	Std Error
Intercept	5.8e−12	1280.863
Residuals Trade Bal (%GDP)	1157.7626	200.8408

Without Luxembourg (n = 29):

Term	Estimate	SE	t stat	p-val
Intercept	−4622.225	4796.003	−0.96	0.3440
Trade Bal (%GDP)	959.60593	232.7805	4.12	0.0003
Muni Waste (kg/person)	62.184369	9.153925	6.79	<.0001

41. **Promotion**

    (a) Yes, as can be seen in the column of timeplots shown in the scatterplot matrix. There is a particularly strong downward trend in the level of sampling (correlation with Week of −0.68). An outlier in the level of detailing during week 6 is also apparent.

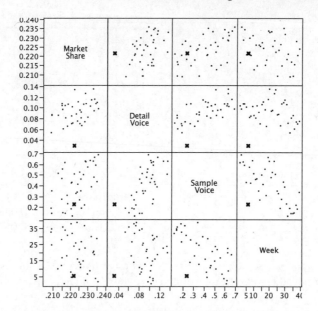

(b) Yes, $F \approx [0.294/(1 - 0.294)] \times (39 - 1 - 3)/3 \approx 4.86 > 4$, so this is a statistically significant amount of variation to explain using a model with $q = 3$ predictors.

(c) Yes. In particular, the marginal effect of detailing (see the scatterplot matrix, with correlation 0.38) is evident and clearly positive. In the multiple regression, the effect is negative (and not statistically significant).

(d) If you find the R^2 from the regression of Detail Voice on Sample Voice and Week, you can calculate that VIF for this variable is VIF(Detail Voice) = $1/(1 - 0.572) \approx 2.34$. Sample voice, in fact has the largest VIF. The R^2 from the regression of these predictor on the others is 0.762, so VIF $\approx 1/(1 - 0.762) \approx 4.20$.

(e) Sampling is the key driver of market share. Though both sampling and detailing are related marginally to the response, detailing only seems useful in the sense that this is the opportunity to hand out samples. Curiously, sampling is also the only one of the 3 variables that contributes statistically significant variation in this fit. VIF only tells you about collinearity among the explanatory variables, not how they are related to the response.

(f) Though not so in this example, the two insignificant variables might be highly correlated with each other. Having one or the other would be useful, just not both.

R^2		0.294436
s_e		0.006632
n		39

Term	Estimate	Std Error	t	p-value	VIF
Intercept	0.2097905	0.005846	35.89	<.0001	.
Detail Voice	−0.011666	0.074983	−0.16	0.8773	2.337
Sample Voice	0.0298901	0.012845	2.33	0.0259	4.193
Week	0.0001239	0.000147	0.84	0.4059	2.436

Chapter 25 Categorical Explanatory Variables

Mix and Match

1. i
3. a
5. b
7. e
9. d

True/False

11. True. It's only possible confounding. The lurking variable must also be related to the response.

13. True

15. True

17. False. The purpose of an interaction is to allow the slopes to differ. Without an interaction, the slopes match.

19. False. We'd have to do this for every conceivable lurking variable, and we've not measured them all. Confounding is always possible without randomization.

21. False. It is helpful if the sizes of the two groups are similar, but not assumed by the model.

23. False. Only four are needed to represent five groups.

Think About It

25. Is this data from a randomized experiment? If not, do we know that the sales agents sell comparable products that produce similar revenue streams? Do we know the costs for the agents in the two groups are comparable, with similar supporting budgets, such as comparable levels of advertising and internal staff support? Without such balance, there are many sources of confounding that could explain the differences that we see in the figure. The lurking variable might also explain the slight difference in variation that we see in the summary.

27. We combine them in order to compare the intercepts and compare the slopes. The multiple regression that combines them include one coefficient that is the difference in the intercepts and another that is the difference between the slopes. These both come with standard errors, and hence allow us to test whether the observed differences (which are the same with either approach) are statistically significant.

29. In general, one should *always* try an interaction unless you have strong reason to know that the slopes are parallel in the two groups. In this context, it seems clear that the model needs an interaction. Union labor in auto plants make more than nonunion labor, and the slope is the estimate for the cost per hour of labor. We'd expect it to be higher in the union shop.

31. (a) The intercept is the mean salary for *Group*=0, namely the women ($140,034). The slope is the difference in salaries, with men marginally making $4,671 more than women overall (*i.e.*, ignoring the effect of managerial grade level).
    (b) These match. The slope in the simple regression is the difference in mean salaries, so regression assigns this estimate almost same level of significance found in the two-sample *t*-test.
    (c) The variances in the two groups are the same. The regression approach is comparable to a two-sample *t*-test that requires equal variances. The *t*-test introduced in Chapter 18 does not require this assumption.

33. (a) About 2. Consider the green points. At $x = 0$, the average seems to be about 0. At $x = 4$, the average of these is near 8. 8/4 = 2. A similar calculation applies to the red points and gives a similar slope near 2.
    (b) The slope will be much flatter, closer to zero. It seems like it might be positive, but will be considerably less than 2.

35. (a) Yes, the fits appear parallel because the coefficient of the interaction ($D\,x$), which measures the difference in the slopes of the two groups, is not statistically significant (its t-statistic is within 1 of zero).
(b) The coefficients of D and $x\,D$ would change in sign, with both becoming positive. Otherwise, the rest of the output would remain as shown.
(c) Remove the interaction term to reduce the collinearity and force the slopes to be precisely parallel.

You Do It

37. **Emerald diamonds**
(a) In order to be a confounding variable, the weight has to be related to the price (we know this is true from previous study of these data, and common sense) and the weight has to be related to the group indicator. That is, diamonds of one clarity grade have to have different weights than those of the other. If the two groups have comparable weights, then the effect of weight is balanced between the two. A two-sample comparison of weight by clarity shows that the average weight is almost the same in the two groups. Weight is unlikely to be a confounding effect in this analysis.

Level	Number	Mean	Std Dev
VS1	90	0.413556	0.054408
VVS1	54	0.408148	0.053661

(b) The two-sample t-test finds a statistically significant difference, with VVS1 costing on average about $112 more than VS1 diamonds.

VS1-VVS1, allowing unequal variances

Difference	−112.30	t Ratio	−2.88504
Std Err Dif	38.93	DF	103.4548
Upper CL Dif	−35.11	Prob > |t|	0.0048
Lower CL Dif	−189.50	Prob > t	0.9976
Confidence	0.95	Prob < t	0.0024

(c) Because the interaction is not statistically significant, we'll remove it and refit the model without this term. Evidently, the costs of either type of diamond rise at the same rate with weight.

Term	Estimate	Std Error	t Ratio	Prob>|t|
Intercept	−52.53705	131.9049	−0.40	0.6910
Weight (carats)	2863.4963	316.2582	9.05	<.0001
Clarity	214.1266	215.9869	0.99	0.3232
Clarity * Weight	−211.5379	522.1921	−0.41	0.6860

Without the interaction, the fits are parallel and the estimated effect for clarity is statistically significant.

R^2	0.497376
s_e	161.8489
n	144

Term	Estimate	Std Error	t Ratio	Prob>|t|
Intercept	−20.44887	105.1595	−0.19	0.8461
Weight (carats)	2785.9054	250.9129	11.10	<.0001
Clarity	127.36823	27.89249	4.57	<.0001

Based on the fit of this multiple regression, we see that for diamonds of comparable weight, those of clarity VVS1 cost on average about $127 more than those of clarity VS1.

(d) From the two-sample comparison, the 95% confidence interval for the mean difference in price is $35 to $190 more for VVS1 diamonds. The estimated mean difference from the multiple regression is

$127.36823 - 2 \times 27.89249, 127.36823 + 2 \times 27.89249 \approx \$72$ to $183

The regression interval is shorter because it removed the variation price due to weight, providing a more precise estimate. There's no confounding however, because the weights are comparable in the two groups. Hence the estimated average differences in price ($112 vs $127) are comparable.

(e) The two groups have similar variances, but the variance increases with the price. Thus, the multiple regression does not meet the similar variances condition. The prices of diamonds become more variable as they get larger.

Chapter 25 Categorical Explanatory Variables

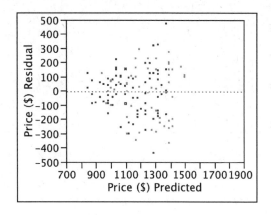

39. **Download**
(a) The file size is related to the transmission time. To be a confounding variable it must also be different in the two groups. As shown in the two sample comparison, the file sizes are paired in the two groups. Because of this balance, the file size cannot be a confounding variable. It's the same in both samples.

Level	Number	Mean	Std Dev
MS	40	56.9500	25.7014
NP	40	56.9500	25.7014

(b) The two-sample t-test finds a very statistically significant difference in the performance of the software from the two vendors. On average, the software labeled "MS" transfers files in about 5.5 fewer seconds. (The variance is substantially larger for the files sent using the NP software.)

MS-NP, allowing unequal variances

Difference	−5.5350	t Ratio	−2.52682
Std Err Dif	2.1905	DF	58.79005
Upper CL Dif	−1.1515	Prob > |t|	0.0142
Lower CL Dif	−9.9185	Prob > t	0.9929
Confidence	0.95	Prob < t	0.0071

(c) The interaction in the model is statistically significant, meaning that the two types of software have different rates of transfer (different MB per second).

R^2	0.752229
s_e	5.138168
n	80

Term	Estimate	Std Error	t Ratio	Prob>|t|
Intercept	4.8929786	1.995934	2.45	0.0165
File Size (MB)	0.4037229	0.032012	12.61	<.0001
Vendor Dummy	4.7633694	2.822677	1.69	0.0956
Vendor Dummy * File Size	−0.180832	0.045272	−3.99	0.0001

As shown in this plot of the fit of this model (different intercepts and slopes in the two groups), the transfer times using MS (darker) become progressively less than obtained by the software labeled NP. The small difference in the intercepts (the coefficient of the dummy variable is not statistically significant) happens because both send small files quickly. The difference emerges only when the files get larger.

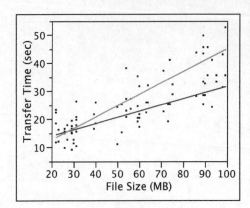

(d) The two-sample comparison finds an average difference of 5.5 seconds (range 1 to 10 seconds), with MS transferring files faster. The analysis of covariance also identifies MS as faster, but shows that the gap becomes progressively wider as the file size increases. NP transfers files (once started) at a rate of about 0.4 sec/MB compared to 0.4 sec/MB for MS. The mean of the two-sample comparison is an average gap ignoring the size of the files.

(e) No. You can see hints of a problem in the color-coded plot of residuals on fitted values (with MS shown in red). Similarly, the boxplots of residuals show different variances.

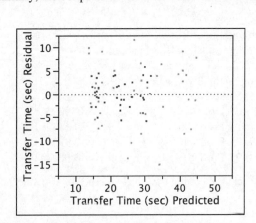
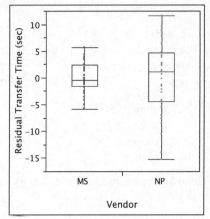

41. **Home prices**
(a) There's a clear difference, with a much steeper slope (higher fixed costs) for the data for Realtor B (shown as crosses here).

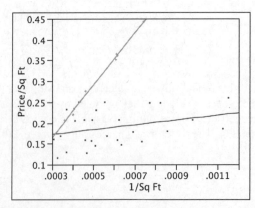

(b) The model requires both a dummy variable and interaction.

R^2	0.762904
s_e	0.037308
n	36

Chapter 25 Categorical Explanatory Variables 95

Term	Estimate	Std Error	t Ratio	Prob>|t|
Intercept	0.155721	0.019713	7.90	<.0001
1/Sq Ft	57.923342	31.2019	1.86	0.0726
Realtor Dummy	−0.176852	0.061062	−2.90	0.0068
Dummy * 1/SqFt	568.5921	110.8419	5.13	<.0001

(c) The data for Realtor B is much less variable around the fitted line than for Realtor A; the residuals do not meet the similar variances condition. You don't need to see the boxplots to see the problem in this example, if you've got the points colored.

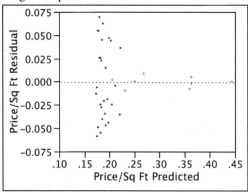
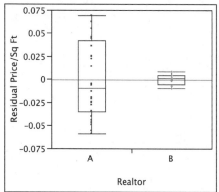

(d) The estimates are fine to interpret even with the evident lack of similar variances, as these reproduce the fitted equations for the separate groups. The intercept, about $156/SqFt is the estimated variable cost for Realtor A homes. The fixed costs for this realtor (slope for 1/SqFt) run about $58,000. For Realtor B, the estimated intercept is near zero (0.158 − 0.177), suggesting no variable costs! Instead, the prices for this realtor seem to be all fixed costs, with an estimate near 58 + 569 = $627,000 regardless of the size of the home.
(e) No, you cannot use these estimates of variation. The formula for the SE of a regression slope depends on the single estimate s_e of residual variation, and that is inappropriate in this analysis. We need to have separate estimates of the variance of for the two realtors. We can interpret the fit, but not use the tools for inference.

43. **R&D expenses**
(a) The two look very similar with the colors evenly mixed. A simple regression to both years seems reasonable.

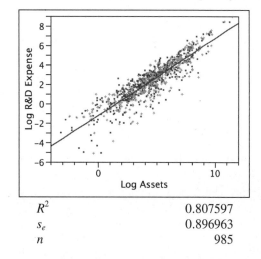

R^2		0.807597
s_e		0.896963
n		985

Term	Estimate	Std Error	t Ratio	Prob>|t|
Intercept	−1.192587	0.062477	−19.09	<.0001
Log Assets	0.7954859	0.012384	64.23	0.0000

(b) The residuals from the multiple regression show some skewness noted previously for 2004. Both have comparable variances, however, and share this problem. As you can tell from the normal quantile plot, the combined data are not nearly normal, but since we are working with the slopes (which are averages) we can continue on by using the central limit theorem. There's a more serious problem, however, not seen in these plots: do you really think that the two data values from AMD or Intel, for example, are independent of each other? Or, does it seem more likely that the data are dependent? We think they are dependent, calling into question any notion of using the usual formulas for standard errors.

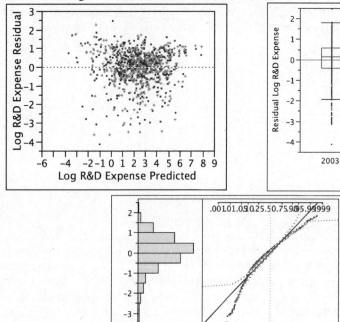

(c) Here's the summary of the multiple regression. Neither added variable is statistically significant and the R^2 has hardly moved from the simple regression.

R^2	0.8077
s_e	0.897636
n	985

Term	Estimate	Std Error	t Ratio	Prob>|t|
Intercept	−1.184021	0.091473	−12.94	<.0001
Log Assets	0.7900453	0.017898	44.14	<.0001
Year Dummy	−0.016828	0.125321	−0.13	0.8932
Dummy * Log Assets	0.0110461	0.024828	0.44	0.6565

The incremental F test that measures the change in R^2 that comes with adding two explanatory variables is
$F = (0.8077 − 0.807597)/(1 − 0.8077) \times (985 − 1 − 3)/2 \approx 0.26$
which is not statistically significant. This agrees with the visual impression conveyed by the original scatterplot: the relationship appears to be the same in both years.

(d) Overall, a common regression model captures the relationship. The elasticity of R&D expenses with respect to assets is about 0.8: on average each 1% increase is assets comes with a 0.8% increase in R&D expenses. There are serious questions, however, about the independence of the residuals in the two years, since there is a pair of measurements on each company. It's hard to think of these as independent.

45. **Movies**
(a) Adult movies (darker dots) appear to have consistently higher subsequent sales at a given box-office gross

Chapter 25 Categorical Explanatory Variables

than family movies. The fits to the two groups look linear (on this log scale) with a "fringe" of outliers. A common simple regression splits the difference between the two groups. Here's the simple regression.

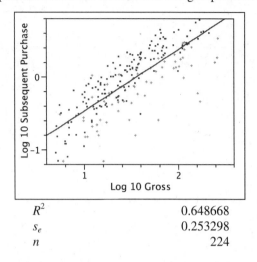

R^2	0.648668
s_e	0.253298
n	224

Term	Estimate	Std Error	t Ratio	Prob>|t|
Intercept	−1.305742	0.063479	−20.57	<.0001
Log 10 Gross	0.8420019	0.04159	20.25	<.0001

(b) The following results summarize fitting the multiple regression with a dummy variable and interaction.

R^2	0.75236
s_e	0.213623
n	224

Term	Estimate	Std Error	t Ratio	Prob>|t|
Intercept	−1.344678	0.104526	−12.86	<.0001
Log 10 Gross	0.7394797	0.063621	11.62	<.0001
Audience Dummy	−0.070228	0.122375	−0.57	0.5666
Dummy * Log Gross	0.2358524	0.076836	3.07	0.0024

The initial scatterplot appears straight enough within groups, and the plot of residuals on fitted values shows no deviations from the conditions. The comparison boxplots show that the variability is consistent in the two groups. The normal quantile plot confirms that the combined residuals are nearly normal, though a bit skewed (toward the "left" and smaller values). A subset of movies (kids' movies, it seems) earns quite a bit less in subsequent sales for their level of box-office success.

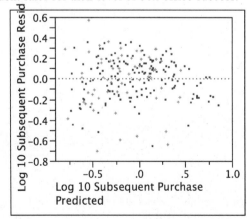
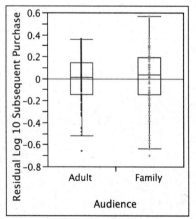

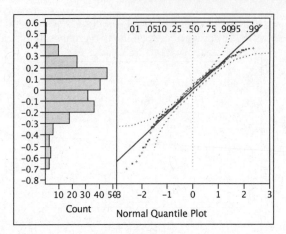

(c) The incremental F-test uses the change in R^2 to measure the statistically significant of adding the two predictors (dummy and interaction). The test statistic is
$$F = (0.75236 - 0.648668)/(1 - 0.75236) \times (224 - 1 - 3)/2 \approx 46$$
which is very statistically significant. We'd reject H_0 that the added variables both have slope zero.
(d) The interaction is highly statistically significant, but the slope for the dummy is not. Only one predictor seems useful. The F-test reaches a more impressive view of the value of adding these two predictors because it is not affected by the substantial collinearity between them. The VIFs for these explanatory variables are almost 15, reducing the size of the shown t-statistic for each by about 4.
(e) The estimates show that as the box-office gross increases, movies intended for adult audiences sell statistically significantly better. Each 1% increase in the box-office gross for an adult movie fetches about 0.74% increase in after-market sales. For adult movies, the elasticity jumps to about 0.74 + 0.24 = 0.98%. As the success at the box office grows, the gap opens up with adult movies doing better.

47. **Promotion**
(a) A simple regression that combines the data from both locations makes a serious mistake, one that vastly overstates the effect/benefit of detailing. By fitting one line to both groups, rather than within each, the higher sales in Boston (darker) inflate the slope.

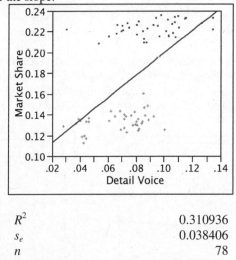

R^2	0.310936
s_e	0.038406
n	78

Term	Estimate	Std Error	t Ratio	Prob>|t|
Intercept	0.0917039	0.015423	5.95	<.0001
Detail Voice	1.0825305	0.184853	5.86	<.0001

(b) The scatterplot suggests using parallel fits in the two locations, with a common slope for detailing. This model meets the MRM conditions. First, we check the interaction and find that it's not statistically significant (the model meets the MRM conditions).

Chapter 25 Categorical Explanatory Variables

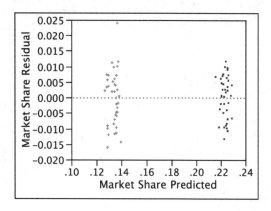

The residuals also do not show substantial tracking over time, with the DW statistic for both being reasonably close to 2. If we plot the residuals from one location on those from the other, there's no association here either.

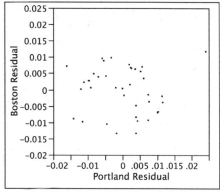

Term	Estimate	Std Error	t Ratio	Prob>|t|
Intercept	0.1230925	0.003255	37.81	<.0001
Detail Voice	0.1521676	0.04406	3.45	0.0009
City Dummy	0.0861738	0.002073	41.57	<.0001

(c) The effect for detailing has fallen from 1.08 down to 0.15, with a range of
    $0.152167 - 2 \times 0.04406, 0.152167 + 2 \times 0.04406 = .064047$ to $.240287$
which rounds to 0.06 to 0.24. Rather than get a 1% gain in market share with each 1% increase in detailing voice, the model estimates a far smaller return on this promotion. By ignoring the effects of the two groups, the analyst inflated the effect of promotion.

Chapter 26 Analysis of Variance

Mix and Match

1. c (This is the two-subscript notation for the response.)
3. h (The fitted value in an ANOVA is the mean of a group.)
5. a (The intercept is the mean of the group not identified by a dummy variable in the multiple regression version of an ANOVA.)
7. g
9. d (The F-statistic uses mean squares, not sums of squares.)

True/False

11. False. Randomization improves any experiment by reducing or eliminating the risk of confounding from a lurking variable
13. True
15. True
17. False. H_0 claims that the groups have the same mean value.
19. True
21. True. Tukey intervals are longer than the corresponding t-intervals in order to adjust for the presence of multiple comparisons, but shorter than Bonferroni adjusted version of the t-intervals

Think About It

23. No. The pooled t-test assumes equal variances in the data, just as in the SRM. The t-test in general does not require this assumption and allows different variances in the two groups; it would produce a different standard error for the difference between the sample means.

25. To change Y from dollars to Euros, you'd divide by 1.5. The sums of squares in the ANOVA table would be smaller by a factor of 1.5^2, keeping the F-statistic unchanged. The change to Euros would divide all of the estimates (intercept and slopes) in the regression by 1.5 as well. For example, the intercept would be $1795.05/1.5 \approx 1197$ Euros and the slope for D(US) would be $1238.10/1.5 \approx 825$. All of the standard errors would also be smaller by a factor of 1.5 as well, so that the t-statistics would remain as shown. Here's the summary in Euros.

Analysis of Variance

Source	DF	Sum of Squares	Mean Square	F Ratio	Prob > F
Regression	3	25809844	8603281	131.1519	<.0001
Residual	96	6297391	65598		
C. Total	99	32107235			

Parameter Estimates

Term	Estimate	Std Error	t Ratio	Prob>|t|
Intercept	1196.7008	51.22415	23.36	<.0001
D(US)	825.39682	72.44188	11.39	<.0001
D(Europe)	478.55877	72.44188	6.61	<.0001
D(Pacific)	−523.7232	72.44188	−7.23	<.0001

27. The results are highly statistically significant. Notice that several groups are almost disjoint, with little overlap between "c" and "b" and "d". The following table summarizes the ANOVA for these data.

Analysis of Variance

Source	DF	Sum of Squares	Mean Square	F statistic	p-value
Regression	4	27835.387	6958.85	9.6449	<.0001

Chapter 26 Analysis of Variance   101

Source	DF	Sum of Squares	Mean Square	F statistic	p-value
Residual	55	39682.648	721.50		

29. The data in two groups determine each slope in a one-way ANOVA, as in the regression view of a two-sample t-test. The least squares regression line has to minimize the deviations away from the line within each group, and that happens if the fit in each group is the mean in that group.

31. Not even close! An analysis of variance fits each group with its own mean, without making assumptions on the relationship among the means. The coding 1, 2, 3, 4, through 5 implies that the means should line up in this order, with equal spacing.

33. We cannot tell; it may or may not include zero. It would be nice if the F-test could tell us this, but it does not. Rejecting H_0 does not imply that any specific confidence interval excludes zero. Rejecting H_0 only means that there's some difference among the means, but it does not indicate which means these are.

35. A Type I error is incorrectly rejecting the null hypothesis (H_0 claims that the difference in mean yields is zero). His chance for a Type I error is much larger than 0.05; he should have adjusted for letting the data pick the pair of sample means to compare and set $\alpha = 0.05/(12 \times 11/2) \approx .00076$ as in the Bonferroni interval.

37. (a) The $J = 54$ varieties implies that the multiple regression requires 53 dummy variables. Quite a chore to create those unless your software does it for you.
(b) Even though n is much larger (8×5 versus 8×54), the Tukey and Bonferroni intervals are quite a bit longer. The analysis of 54 varieties means that we can consider $_{54}C_2 = 54(54 - 1)/2 = 1,431$ pairwise comparisons. The Tukey interval adjusts for multiplicity by increasing the multiplier in the confidence interval from 2.88 up to 4.06. Similarly, the Bonferroni adjustment requires the t critical value for $\alpha = 0.05/1431 = .000035$. That's about 4.19, slightly bigger than the Tukey interval.

You Do It

39. (a) $J = 4$ groups, leaving $60 - 4 = 56$ degrees of freedom for the residuals. The 4 groups require 3 regression coefficients, so the DF for the Between Sum of Squares is 3. The sums of squares add up to the total. The mean squares are ratios of sums of squares divided by degrees of freedom.

Source	DF	Sum of Squares	Mean Square	F	p-value
Regression	3	150	50.000	3.5	0.0212
Residual	56	800	14.286		
Total	59	1000			

(b) The null hypothesis states that the means of the populations defined by this experiment are all equal, $H_0: \mu_1 = \mu_2 = \mu_3 = \mu_4$. In other words, that there's no preference for one sweetener over another.
(c) Yes, because the p-value of the F-statistic is less than 0.05. There is a statistically significant difference among the 4 means.
(d) We can see that there is a statistically significant difference between the average ratings; the type of sweetener affects the ratings. Without the means, we cannot tell which is preferred, but we can see that a difference does exist.

41. (a) Use the Bonferroni procedure to determine the length of the 95% confidence interval when considering comparisons between $_{51}C_2 = 51 \times 50/2 = 1275$ pairs. That vast number of comparisons pushes α down to $0.05/1275 \approx .0000392$. You'll need software to find this percentile of the t-distribution. (The table of the normal distribution gets you in the right ballpark because in this example there are $51 \times (20 - 1) = 969$ degrees of freedom.) The appropriate t-critical value is $t \approx 4.13$. (Most software will not produce percentiles of the distribution needed for Tukey intervals; JMP gives 4.02.) Hence, to be statistically significant, the difference between two sample averages must be at least
$$4.13 \times s_e \sqrt{2/20} = 4.13 \times 3500 \times \sqrt{1/10} \approx \$4,571$$
(b) No, not unless there is some suspicion that some state is out of line. The difference needed to identify a statistically significant difference is quite large compared to the stated belief that most sales are within $2,000.

43. (a) It helps to answer the following questions to think about the plots first. The content of the scatterplot of Y on X is the same as that of Y on D, only with a different scale on the x-axis. The x-axis in the plot of Y on D goes from 0 to 1, whereas it goes from -1 to 1 in the plot of Y on X.

(b) The explanatory X (coded as -1 and $+1$) is perfectly correlated with the dummy variable D; $X = 2D - 1$. Hence, the regression models have the same R^2.
(c) With either explanatory variable, the sample means of the two groups are the fitted values.
(d) The intercept b_0 in the regression of Y on D is the mean for women ($D = 0$) and the slope b_1 is the difference (mean of men minus mean of women). In the regression of Y on X, the intercept is the overall mean value ($b_0 = \bar{y}$); the slope is half of the difference between the means since the two groups are 2 units apart rather than 1.

45. (a) The data do not meet the straight-enough condition. The means in the data appear to first rise (no pun intended) then fall with the amount of potassium bromate.

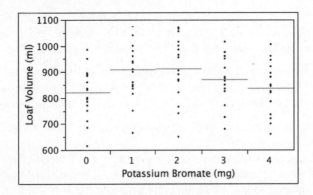

(b) An ANOVA does not require the straight-enough condition, so that is not an issue. The variation in the data, however, seems to increase slightly with the mean value as shown in this summary table, but the effect is not substantial. There are no evident outliers nor extreme skewness, though we need to see the residuals from the ANOVA to check the normal quantile plot.

Pot. Bromate	Number	Mean Volume	Std Dev
0	17	819.118	95.855
1	17	907.941	111.524
2	17	910.882	120.341
3	17	869.412	99.889
4	17	835.882	96.570

(c) Yes, the bakery can manipulate the volume by varying this ingredient. The F-statistic is barely statistically significant ($p = 0.0404 < 0.05$). We can reject the null hypothesis and conclude that there is a statistically significant difference in loaf volumes.

Analysis of Variance

Source	DF	Sum of Squares	Mean Square	F	p
Regression	4	116537.1	29134.3	2.6289	0.0404
Residual	80	886582.4	11082.3		
Total	84	1003119.4			

(d) The estimated standard error of the difference is
$$s_e \sqrt{2/n_j} = \sqrt{s_e^2 \, 2/n_j} = \sqrt{11082.3 \times 2/17} \approx 36.1 \text{ ml}$$
The Tukey interval is
$$\bar{y}_2 - \bar{y}_0 \pm 2.79 \times s_e \sqrt{2/n_j} = 910.882 - 819.118 \pm 2.79 \times 36.1$$
$$= 91.764 \text{ ml} \pm 100.72$$
The difference is not statistically significant.
(e) The Bonferroni interval uses the same standard error as the Tukey interval, but with a t-critical value using $85 - 5 = 80$ degrees of freedom and $\alpha = 0.05/(5 \times 4/2) = 0.005$. Hence the interval is
$$\bar{y}_2 - \bar{y}_0 \pm t \times s_e \sqrt{2/n_j} = 910.882 - 819.118 \pm 2.89 \times 36.1$$
$$= 91.764 \text{ ml} \pm 104.329$$
The difference is not statistically significant.
(f) The F-test rejects H_0: $\mu_1 = \mu_2 = \mu_3 = \mu_4 = \mu_5$. The Tukey and Bonferroni intervals, however, find that the largest pairwise difference (between 0 mg and 2 mg) is not statistically significant. Evidently, whatever

difference exists among the means is subtle. Perhaps it's the difference between a combination of 1 and 2 mg versus 0 and 4 mg. We cannot tell from this analysis.

47. **Stopping distances**

(a) The plot of the stopping distances organized by type of car makes it clear that there are large, consistent differences among these cars. All 10 stops for the Malibu, for example, are less than the stopping distances of the others.

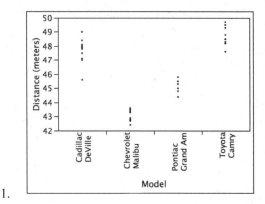

(b) The estimated intercept is the mean stopping distance for the omitted category, here the Toyota Camry (48.66 meters). The slopes for the other models are differences in the group averages. For example, the slope for the Cadillac dummy (−1.02) indicates that the average stopping distance of the Cadillac (47.64 meters) is 1.02 meters less than the average stopping distance for the Camry (48.66 meters). The slope for the Malibu dummy (−5.60) indicates that its stops on average were 5.6 meters shorter than those of the Camry.

(c) The ANOVA table for the multiple regression gives the F-statistic $F = 152$ with p-value far less than 0.05. As suggested by the initial plot, the average differences are highly statistically significant. The variance between groups is 152 times that within the groups.

Analysis of Variance

Source	DF	Sum of Squares	Mean Square	F Ratio	Prob > F
Regression	3	191.659	63.886	151.5890	<.0001
Residual	36	15.172	0.421		
Total	39	206.831			

(d) Changing from meters to feet would increase the estimates in the multiple regression (b_0, b_1, etc.) by a factor of a bit more than 3 (1 meter ≈ 3.28 feet) since these would be averages in feet rather than meters. The overall F-statistic and p-value would be the same.

(e) Yes, but there are serious concerns that require follow-up. The following plot shows the residuals from the multiple regression, grouped by model. The Cadillac stopped rather quickly once (about 2 meters less than average), producing the most visible outlier. The variation is similar in the groups, with a bit more consistency for the Malibu. The normal quantile plot shows that the data are nearly normal. Based on these plots of residuals these data meet the conditions for an ANOVA.

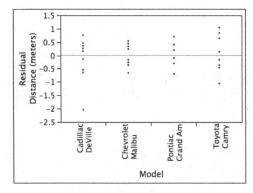 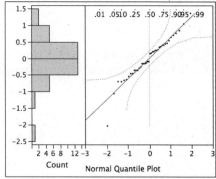

More serious concerns, however, are not revealed in plots. For example, there may be dependence among the measurements. Since the testers repeatedly stopped the same car, had they not taken care to let the brakes cool between stops, the data for the cars may be dependent. (The braking distance for a model would grow longer as the testing proceeded.) Also, there may be confounding. For example, were these cars driven by different people? If so, the data may then compare the 4 drivers rather than the 4 cars.

(f) The test has a "sample" of one car from each model. We cannot infer that all Malibu's stop as well as this one because we have not observed the variation among Malibu's. We can only conclude, assuming the issues raised in (e) are resolved, that there are differences among these 4 cars - not other cars of the same model.

49. **Movie ratings**

(a) Yes, the data appear suited to ANOVA. The variation (judging from the interquartile ranges, lengths of the boxes in this plot) appears similar in the 3 groups.

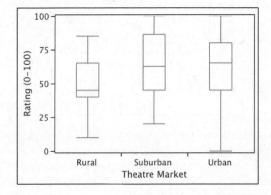

(b) The side-by-side boxplots suggest that the average urban and suburban customer assign comparable ratings. The rural ratings appear smaller. There is considerable variation within the groups compared to the differences among the centers (this plot shows medians, not means). That said, these are large groups and the standard error of the average is small for a large sample.

(c) The intercept is the average rural rating (49.13). The slope for the dummy variable representing urban views (13.70) is the difference between the average rating of urban views and the average rating of rural viewers. (Urban viewers assign higher average ratings.) Similarly, the slope for the dummy variable representing suburban views (15.87) is the difference between the average rating of suburban views and the average rating of rural viewers.

Analysis of Variance

Source	DF	Sum of Squares	Mean Square	F Ratio	Prob > F
Regression	2	4214.73	2107.37	3.9455	0.0202
Residual	344	183737.00	534.12		
Total	346	187951.73			

Parameter Estimates

Term	Estimate	SE	t	Prob>|t|
Intercept	49.13	4.82	10.20	<.0001
Dummy(Urban)	13.70	5.00	2.74	0.0065
Dummy(Suburban)	15.87	6.89	2.30	0.0219

(d) The standard errors differ because the groups have different sample sizes. There are more urban viewers so this SE is smaller than that which compares suburban to rural viewers. For b_1,

$$se(b_1) = se(\bar{y}_u - \bar{y}_r) = s_e \sqrt{1/n_u + 1/n_r} \approx 23.11 \times \sqrt{1/302 + 1/23} \approx 5.00$$

whereas for b_2,

$$se(b_2) = se(\bar{y}_s - \bar{y}_r) = s_e \sqrt{1/n_s + 1/n_r} \approx 23.11 \times \sqrt{1/22 + 1/23} \approx 6.89$$

Though not shown, the standard error for the other comparison (needed in part g) is

$$se(\bar{y}_s - \bar{y}_u) = s_e \sqrt{1/n_s + 1/n_u} \approx 23.11 \times \sqrt{1/22 + 1/302} \approx 5.10$$

Chapter 26 Analysis of Variance

Level	Number	Mean
Rural	23	49.1304
Suburban	22	65.0000
Urban	302	62.8311

(e) The differences among the group means are statistically significant. The *p*-value for the *F*-test is 0.0202 < 0.05. Reject H_0 that the means in the population are the same. The relatively small value of the *F*-statistic indicates that, as seen in the plot, the differences between the means are not so large relative to the variation within the groups.

(f) The boxplots of the residuals and normal quantile plot of the residuals do not indicate a problem. The urban boxplot has longer whiskers because there are more people in this group. The size of the box itself is rather similar to those of the others. The residuals are nearly normal, but there's quite a pile of them at the high side; it appears that the data are truncated at the high end (people would have scored some movies "more than 100" if they could have).

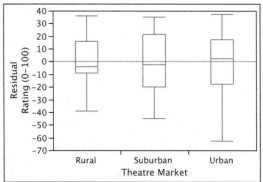
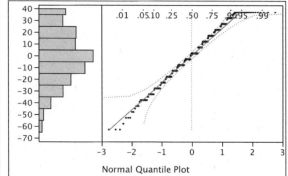

We might ask about whether the viewers interacted while watching the movie, however, as that might produce dependence. (That's pretty likely, for example, if they watched the movie at the same time together.)

(g) The film gets lower ratings in the rural locations, but there is substantial variation. The only statistically significant difference is between the mean rating in the urban district differs from the mean rating in the rural district. To reach this conclusion, we used software to compute Tukey confidence intervals. (The same comments apply to Bonferroni intervals that are slightly longer). Even though the suburban viewers assign the highest rating, the mean of suburban viewers is not statistically significantly larger than that of rural viewers, even though the mean of urban viewers is. This occurs because of the large number of urban viewers. We have enough urban viewers to distinguish their mean difference from rural viewers (13.7), but not enough to distinguish suburban viewers (15.87).

Level	Mean
Suburban	65.00
Urban	62.83
Rural	49.13

Level	– Level	Difference	Lower CL	Upper CL
Suburban	Rural	15.86957	−0.35372	32.09285
Urban	Rural	13.70069	1.93325	25.46813
Suburban	Urban	2.16887	−9.84451	14.18226

For Bonferroni intervals, the calculations are similar with slightly longer intervals. The *t*-statistic using $\alpha = 0.05/3 \approx 2.41$

Level	– Level	Difference	SE	Lower CL	Upper CL
Suburban	Rural	15.87	6.89	−0.74	32.47
Urban	Rural	13.70	5.00	1.65	25.75
Suburban	Urban	2.17	5.10	−10.12	14.46

Chapter 27 Time Series

Mix and Match

1. d

3. a

5. b

7. e

9. f

True/False

11. True

13. True. A moving average uses data both before and after the location of the smoothed value; an exponential smooth is one-sided.

15. True. An exponential smooth is an average, and outliers affect all averages.

17. False. The accuracy is low near the edges of the data and can fall off dramatically as we extrapolate.

19. False. See the example of this chapter.

21. True. You can still use it as a short-cut method to find the first autocorrelation of the residuals, but you cannot rely on the D statistic in this situation

Think About It

23. (a) A 5–term average in general is smoother because it averages more data. It's a lot like the mean: the more data we average, the less variation from average to average.
    (b) We don't know values for the future of the series, y_{n+1}, y_{n+2} and so on. So, there's less averaging that can be done. The exponentially weighted moving average has problems getting started, but it's working fine at the end of the time series.
    (c) The time series is no longer centered at the same time points as the data values. For example, suppose that we use a four-term moving average. Should it be defined as
    $$\bar{y}_{t,4} = \frac{y_{t-2} + y_{t-1} + y_t + y_{t+1}}{4} \quad \text{or as} \quad \bar{y}_{t,4} = \frac{y_{t-1} + y_t + y_{t+1} + y_{t-2}}{4} ?$$

25. (a) The forecast is that the series will stay at the mean: $50 + 0.8 \times 250 = \$250$ thousand.
    (b) Return toward the mean. The prediction is $50 + 0.8 \times 300 = \$290$ thousand.
    (c) The forecast is for an increase whenever the current value is smaller than the mean, and for a decrease whenever it's greater than the mean. The process is said to be "mean-reverting."

27. (a) The EMWA with less smoothing, $w = 0.5$. The other smoother overly smoothes the data.
    (b) We'd guess growth to be about -0.5 percent in the next quarter, roughly the location of the end of the EWMA. That puts GNP at $14.238 \times (1 - 0.005) \approx \$14.167$ trillion
    (c) From the variation around the EWMA, we'd guess uncertainty of about $\pm 1\%$ or more. The range for GDP is then 0.99×14.167, $1.01 \times 14.167 \approx \$14.025$ to $\$14.309$ trillion

29. (a) The estimated correlation is weak and negative at $r = -0.12$.
    (b) The one outlier in the time series shows up as the two separated points in the lower left corner of the scatterplot. Because the scatterplot effectively uses each value of y_t twice (once on the vertical axis, and then once on the horizontal axis), an outlier in a time series appears as two points in the scatterplot of y_t on y_{t-1}.
    (c) Yes. The pair of points associated with the outlier produce a positive direction to the correlation, whereas the main cluster has a negative slope. Without the pair from the outlier, the correlation falls to -0.26.
    (d) One has to exclude two points from the scatterplot.

Chapter 27 Time Series

31. (a) Recall that our data begin long before 2000, but we are only using data for the more recent time period. We have prior values to find the lags, so we do not lose the first observation.

    (b) The slope is smaller by one, the intercept is the same. The slope changes because we've moved most of the lag into the response. Algebraically, the equation with changes is exactly the same as the equation with levels.
    $$\hat{y}_t = 0.9 + 0.97 y_{t-1} \rightarrow y_t - y_{t-1} = 0.7 - 0.03 y_{t-1}$$

    (c) The SD of the residuals s_e is the same because the residuals are the same. A residual in the model for shipments y_t is $e_t = y_t - \hat{y}_t = y_t - (0.9 + 0.97 y_{t-1}) = (y_t - y_{t-1}) - (0.9 - 0.03 y_{t-1})$ which is a residual in the model for the changes in shipments.

    (d) Most of the structure explained by the autoregression in the text is contained in the proximity of y_t to y_{t-1}. By differencing the response, we've removed the easy part of the forecast. This model for the differences has to explain the change in the series, and that's much harder. The t-statistic for the slope in the regression of y_t on y_{t-1} compares the estimate to 0; in a sense, the t-statistic in the regression of changes on y_{t-1} is comparing the slope to 1 and is telling us that the estimate is not far from 1.

You Do It

33. **Exxon**

    (a) The time trend, Estimated Exxon Price = 74.803157 + 1.518671 Month,
    misses considerable pattern in the prices, as you can see meandering patterns around the estimated line. The Durbin-Watson statistic $D = 0.15$, implying substantial autocorrelation in the residuals. This is also evident in the timeplot of the residuals at the right below.

    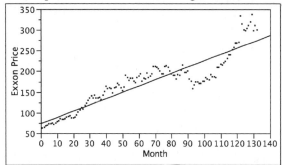 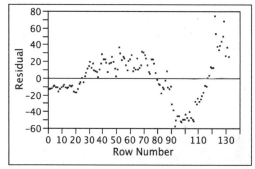

    (b) The fit of the model is summarized in this table. Both lags do not seem needed. The t-statistic of the second lag is not too far from zero, suggesting we can get a more parsimonious model with less collinearity by removing this term

RSquare	0.974062
Root Mean Square Error	10.21448
Observations (or Sum Wgts)	130

Term	Estimate	Std Error	t Ratio	Prob>|t|
Intercept	6.40934	3.16132	2.03	0.0447
Month	0.10069	0.05761	1.75	0.0829
Lag 1 Price	0.79758	0.08853	9.01	<.0001
Lag 2 Price	0.13910	0.08916	1.56	0.1212

    (c) The following results summarize the model with a time trend and one lag of the response. The slope of the time trend has fallen from near 1.5 in the original simple regression to about 0.1. The change in the slope is due to collinearity. The prices themselves are rising linearly, as seen in the model fit in (a).

RSquare	0.974204
Root Mean Square Error	10.23189
Observations (or Sum Wgts)	131

Term	Estimate	Std Error	t Ratio	Prob>|t|
Intercept	7.0647425	3.105349	2.28	0.0246
Month	0.1171149	0.056729	2.06	0.0410
Lag 1 Price	0.9251095	0.034157	27.08	<.0001

(d) No. Although the Durbin-Watson statistic is close to 2, the residuals do not have similar variances, as can be seen in the plot of residuals on fitted values. The prices become more variable as they get larger. We also have a big outlier (February 2005) when the price moved up more than expected.

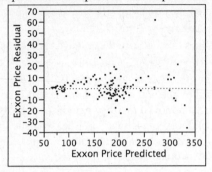

(e) The similarly formulated model using returns finds nothing. The returns on Exxon during this period are evidently simple and models of this type do not identify a pattern.

RSquare		0.016168
Root Mean Square Error		0.050128
Observations (or Sum Wgts)		131

Term	Estimate	Std Error	t Ratio	Prob>|t|
Intercept	0.0230028	0.009098	2.53	0.0127
Month	−0.000126	0.000116	−1.09	0.2791
Lag 1 Return	−0.092142	0.088205	−1.04	0.2982

(f) We'd use the returns and guess that the return next month would look like prior returns…averaging about 1.3% increase per month. The distribution of the returns is nearly normal, (with a bit of fat tails – kurtosis – on the high side), so we could use the SD of the returns (5% per month) to quantify our uncertainty.

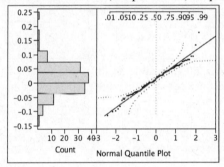

35. **Compensation**
(a) The residuals clearly contain a pattern ($D = 0.05$). The pattern is present because we fit the wrong model and missed the bend in the trend. It's not helpful to describe this structure in the residuals as autocorrelation because that does not help recognize the way to fix the problem: namely, fit a model that captures the trend more completely.

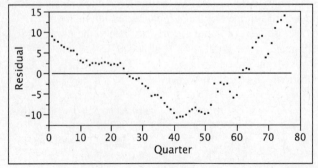

(b) The following output summarizes the fit of the multiple regression.

RSquare		0.995608
Root Mean Square Error		1.919356
Observations (or Sum Wgts)		77

Term	Estimate	Std Error	t Ratio	Prob>|t|
Intercept	79.457816	0.603064	131.76	<.0001
Quarter	0.8109698	0.024434	33.19	<.0001
Late Dummy	−54.40359	2.04544	−26.60	<.0001
Late × Quarter	1.1756546	0.04036	29.13	<.0001

To interpret the fit of the model, start with Dummy = 0, the baseline period (first 42 quarters). For these, the data produce a linear trend

Estimated Hourly Comp = 79.5 + 0.811 *Quarter*

In the second period (with Dummy = 1), the fit includes the dummy slope (changing the intercept) and the interaction (changing the slope) and becomes

Estimated Hourly Comp = (79.5 − 54.4) + (0.81+ 1.18) *Quarter* = 25.1 + 1.98 *Quarter*

Basically, the multiple regression combines two linear regressions, one that fits the early trend, and the second that fits the later linear trend. This plot shows the "segmented" nature of the fitted model.

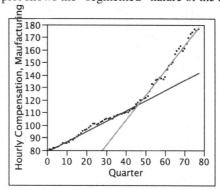

(c) No, the model (while improving the linear trend) does not meet the MRM conditions. A sequence plot of the residuals shows that the model leaves substantial pattern in the residuals, and the Durbin-Watson statistic $D = 0.66$ finds statistically significant autocorrelation.

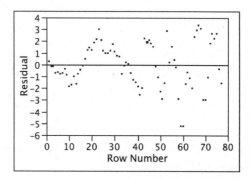

(d) The percentage changes fluctuate around their mean value with no evident pattern or dependence. The variation increased in 2000 (Quarter 53) and appears to have remained large.

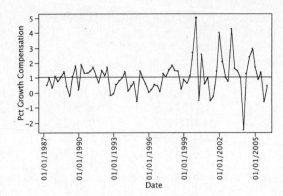

(e) Keep it simple. We'd use the percentage changes and estimate the variation using the data since 2000. The SD in the quarters since 2000 is about 1.65 and the changes are nearly normal (though we only have 25 quarters in this period). The mean seems reasonably stable, so we'd estimated this at the overall mean, about 1 percent per quarter. We'd predict the next value to be 1% larger than last observation in the series (176.44 × 1.01), with interval determined by the SD of the percentage changes:

$$176.466 \times (1.01 - 2 \times 0.0165) = 172.407$$

to

$$176.466 \times (1.01 + 2 \times 0.0165) = 184.054$$

37. Inventory

(a) The fit of the equation is rather impressive, with very large R^2 and highly statistically significant estimated slopes. The intercept is difficult to interpret directly as an estimate at 0 because that is too far from the data. Its relevance is only in the sense of producing an estimated inventory level for these levels of *Debt*. To interpret the slopes, it may be helpful to write the equation slightly differently as

Estimated Inventory = $-6500 - 19 (Debt_{t-1} - Debt_{t-3}) + 5\ Debt_{t-3}$

When Debt goes up, the inventory goes down. High levels of consumer debt in the past (3 quarters ago) are associated with larger inventory levels at Wal-Mart. While it may be tempting to think of a causal justification for this, remember that all of these series are increasing over time. Debt has risen over time along with inventories.

	RSquare		0.966557	
	Root Mean Square Error		1906.756	
	Observations (or Sum Wgts)		101	

Term	Estimate	Std Error	t Ratio	Prob>|t|
Intercept	−6461.261	415.2989	−15.56	<.0001
Lag 1 Debt	−19.22679	3.75513	−5.12	<.0001
Lag 3 Debt	24.327444	3.950198	6.16	<.0001

(b) Leave it out. Adding the second lag introduces more collinearity and obfuscates the form of the model without adding to the R^2.

	RSquare		0.966704	
	Root Mean Square Error		1912.355	
	Observations (or Sum Wgts)		101	

Term	Estimate	Std Error	t Ratio	Prob>|t|
Intercept	−6451.735	416.7737	−15.48	<.0001
Lag 1 Debt	−13.1704	10.00503	−1.32	0.1911
Lag 3 Debt	30.783376	10.64528	2.89	0.0047
Lag 2 Debt	−12.51032	19.14664	−0.65	0.5150

(c) No, it does not. Though very statistically significant, the residuals have a great deal of structure and are not simple. This timeplot shows the residuals, with color-coding and symbols to distinguish the quarters. The data are also autocorrelated within each quarter.

Chapter 27 Time Series

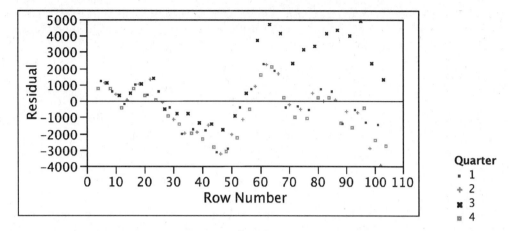

(d) The prior plot shows the distinction among the quarters. Inventory levels are larger than expected in Q3, as if Wal-Mart is loading up for the holiday selling season.

(e) The model would look to get Q3 correct (by accident) and over-predict the inventory levels in the other 3 quarters. Think of the horizontal line in the previous timeplot as the forecasts from the model. Continuing this line to the left and looking at the recent patterns in the residuals in each quarter, it appears that Q3 will fall into place. It also seems that the model will over-predict the other 3 quarters, which have been tracking below the estimates.

39. (a) The timeplot shows the dramatic collapse of the housing market associated with problematic home loans in 2007-2008. At the point of the drop, it seems clear that permits leads the way. Other than for this dramatic drop, it is difficult to separate changes in one from the other. (Permits are lighter, homes completed darker.)

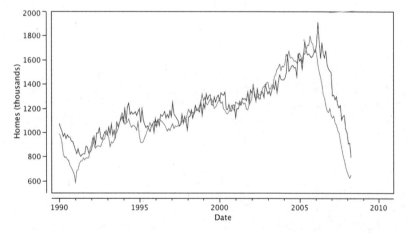

(b) The contemporaneous correlation between completions and permits is 0.888 whereas it is larger (0.967) when using the lagged number of permits. Plots confirm that the lagged number is a better predictor. (Your correlations may differ slightly from these, depending on how you handle the presence of missing values introduced by lags.)

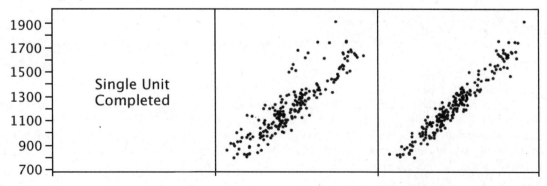

(c) A time trend would have provided a nice fit prior to the collapse of the housing market, but is less useful after the collapse. We found the following model using lags of permits and one lag of housing itself (conveniently 3 months lagged to work with the prediction) to be a good fit.

R^2	0.9497
s_e	51.9587
n	211.0000

Term	Estimate	Std Error	t Ratio	Prob>|t|
Intercept	154.8836	22.1732	6.99	<.0001
Lag 3 Completed	0.1931	0.0636	3.04	0.0027
Lag 3 Permits	0.2940	0.0512	5.74	<.0001
Lag 6 Permits	0.2299	0.0817	2.81	0.0054
Lag 9 Permits	0.1865	0.0660	2.83	0.0052

The timeplot of the residuals shows the presence of a slight amount of autocorrelation in the residuals. Note the slightly bowed appearance suggested by the superimposed red curve in this timeplot.

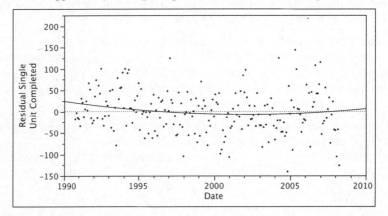

We could remedy this by adding a second-degree polynomial to the regression, but we think that this adjustment is unlikely to improve the predictions. (The fit is modestly better, with R^2 growing slightly.) A greater source of concern is that the variance of the residuals may be increasing over time, suggesting that the model will be less accurate during the turbulent housing market near the end of the data. Otherwise, the plot of the residuals on the fitted values shows consistent variation. The residuals also appear normally distributed.

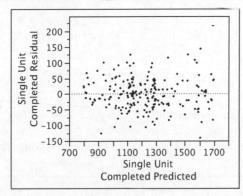
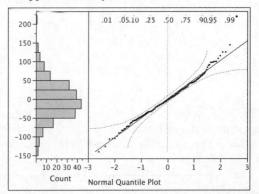

(d) We substituted the values of the lagged variables into our regression and then let the software build the forecast. The equation of the model estimates housing to be

$154.8836 + 0.1931\ y_{t-3} + 0.294\ x_{t-3} + 0.2299\ x_{t-6} + 0.1865\ x_{t-9}$
$= 154.8836 + 0.1931 \times 792 + 0.294 \times 649 + 0.2299\ 675 + 0.1865 \times 811 \approx 805{,}000$ homes

Our software reported the prediction interval as 700,000 to 910,000, close to the approximate prediction interval (which uses $\pm 2s_e$). Our concern with this forecast is a suspicion that the housing market has undergone a dramatic change with the sudden decrease that began with the problems in the mortgage market. The patterns that held in the past may not continue, and the hint that the residuals are becoming more variable (see the time plot above) is a further pointer in this direction.